P9-EKY-884

Environmental Issues

AIR QUALITY

Environmental Issues

AIR QUALITY
CLIMATE CHANGE
CONSERVATION
ENVIRONMENTAL POLICY
WATER POLLUTION
WILDLIFE PROTECTION

Environmental Issues

AIR QUALITY

Yael Calhoun
Series Editor

Foreword by David Seideman,
Editor-in-Chief, *Audubon* Magazine

CHELSEA HOUSE
PUBLISHERS
A Haights Cross Communications Company ®
Philadelphia

CHELSEA HOUSE PUBLISHERS
VP, NEW PRODUCT DEVELOPMENT   Sally Cheney
DIRECTOR OF PRODUCTION   Kim Shinners
CREATIVE MANAGER   Takeshi Takahashi
MANUFACTURING MANAGER   Diann Grasse

Staff for AIR QUALITY
EXECUTIVE EDITOR   Tara Koellhoffer
EDITORIAL ASSISTANT   Kuorkor Dzani
PRODUCTION EDITOR   Noelle Nardone
PHOTO EDITOR   Sarah Bloom
SERIES AND COVER DESIGNER   Keith Trego
LAYOUT   21st Century Publishing and Communications, Inc.

A Haights Cross Communications ⚓ Company ®

First Printing

9 8 7 6 5 4 3 2 1

Library of Congress Cataloging-in-Publication Data

Air quality/[edited by Yael Calhoun]; foreword by David Seideman,
    p. cm.—(Environmental issues)
      Includes bibliographical references and index.
    ISBN 0-7910-8201-6
  1. Air—Pollution. 2. Air quality management—United States. I. Calhoun,
Yael. II. Series.
TD883.A4784 2005
363.739'2—dc22

                                            2004029003

All links and web addresses were checked and verified to be correct at the time of
publication. Because of the dynamic nature of the web, some addresses and links
may have changed since publication and may no longer be valid.

Contents Overview

Foreword by David Seideman, Editor-in-Chief, *Audubon* Magazine   viii

Introduction: "Why Should We Care?"   xiv

Section A:
Air Quality Issues and Challenges   1

Section B:
Health Issues   29

Section C:
Pollution From Transportation and Industry   47

Section D:
Global Dust   107

Bibliography   119

Further Reading   120

Index   121

Detailed Table of Contents

Foreword by David Seideman, Editor-in-Chief, *Audubon* Magazine   viii

Introduction: "Why Should We Care?"   xiv

Section A:

Air Quality Issues and Challenges   1

How Clean Is Our Air?   2

State of the Air 2004   3
from the American Lung Association

What Pollutes the Air We Breathe?   11

Air Pollutant Factsheets   12
from the U.S. Environmental Protection Agency

Section B:

Health Issues   29

Is Air Pollution a Factor in Heart Disease?   30

The Effects of Air Pollution on Health   31
from the American Heart Association

Does Air Pollution Contribute to Asthma and Respiratory Problems?   37

Is Air Pollution Making Us Sick?   37
by Kimi Eisele

Section C:

Pollution From Transportation and Industry   47

What Do Diesel Engines Contribute to Air Pollution?   48

Closing the Diesel Divide   49
by Hillary Decker et al.

How Has the Clean Air Act Helped Clean Our Air?   63

Building on Thirty Years of Clean Air Success   64
by Elissa Gutt et al.

Is Air Pollution a New Problem?   74

    **Earth Under Siege: From Air Pollution to
        Global Change**   75
    by Richard P. Turco

Is Acid Rain Still an Environmental Issue?   85

    **Unfinished Business: Why the Acid Rain
        Problem Is Not Solved**   86
    from the Clean Air Task Force

**What Is the Problem With Mercury From
Power Plants?**   96

    **Out of Control and Close to Home: Mercury
        Pollution from Power Plants**   97.
    by Michael Shore

Section D:
Global Dust   107

**How Does Dust Pollute Air and Water
Around the Globe?**   108

    **Dust in the Wind: Fallout From Africa May
        Be Killing Coral Reefs an Ocean Away**   109
    by John C. Ryan

    Dusted   114
    by Hannah Holmes

Bibliography   119
Further Reading   120
Index   121

FOREWORD

by David Seideman, Editor-in-Chief, *Audubon* Magazine

For anyone contemplating the Earth's fate, there's probably no more instructive case study than the Florida Everglades. When European explorers first arrived there in the mid-1800s, they discovered a lush, tropical wilderness with dense sawgrass, marshes, mangrove forests, lakes, and tree islands. By the early 20th century, developers and politicians had begun building a series of canals and dikes to siphon off the region's water. They succeeded in creating an agricultural and real estate boom, and to some degree, they offset floods and droughts. But the ecological cost was exorbitant. Today, half of the Everglades' wetlands have been lost, its water is polluted by runoff from farms, and much of its wildlife, including Florida panthers and many wading birds such as wood storks, are hanging on by a thread.

Yet there has been a renewed sense of hope in the Everglades since 2001, when the state of Florida and the federal government approved a comprehensive $7.8 billion restoration plan, the biggest recovery of its kind in history. During the next four decades, ecologists and engineers will work to undo years of ecological damage by redirecting water back into the Everglades' dried-up marshes. "The Everglades are a test," says Joe Podger, an environmentalist. "If we pass, we get to keep the planet."

In fact, as this comprehensive series on environmental issues shows, humankind faces a host of tests that will determine whether we get to keep the planet. The world's crises—air and water pollution, the extinction of species, and climate change—are worsening by the day. The solutions—and there are many practical ones—all demand an extreme sense of urgency. E. O. Wilson, the noted Harvard zoologist, contends that "the world environment is changing so fast that there is a window of opportunity that will close in as little time as the next two or three decades." While Wilson's main concern is the rapid loss of biodiversity, he could have just as easily been discussing climate change or wetlands destruction.

The Earth is suffering the most massive extinction of species since the die-off of dinosaurs 65 million years ago. "If we continue at the current rate of deforestation and destruction of major ecosystems like rain forests and coral reefs, where most of the biodiversity is concentrated," Wilson says, "we will surely lose more than half of all the species of plants and animals on Earth by the end of the 21st century."

Many conservationists still mourn the loss of the passenger pigeon, which, as recently as the late 1800s, flew in miles-long flocks so dense they blocked the sun, turning noontime into nighttime. By 1914, target shooters and market hunters had reduced the species to a single individual, Martha, who lived at the Cincinnati Zoo until, as Peter Matthiessen wrote in *Wildlife in America,* "she blinked for the last time." Despite U.S. laws in place to avert other species from going the way of the passenger pigeon, the latest news is still alarming. In its 2004 State of the Birds report, Audubon noted that 70% of grassland bird species and 36% of shrubland bird species are suffering significant declines. Like the proverbial canary in the coalmine, birds serve as indicators, sounding the alarm about impending threats to environmental and human health.

Besides being an unmitigated moral tragedy, the disappearance of species has profound practical implications. Ninety percent of the world's food production now comes from about a dozen species of plants and eight species of livestock. Geneticists rely on wild populations to replenish varieties of domestic corn, wheat, and other crops, and to boost yields and resistance to disease. "Nature is a natural pharmacopoeia, and new drugs and medicines are being discovered in the wild all the time," wrote Niles Eldredge of the American Museum of Natural History, a noted author on the subject of extinction. "Aspirin comes from the bark of willow trees. Penicillin comes from a mold, a type of fungus." Furthermore, having a wide array of plants and animals improves a region's capacity to cleanse water, enrich soil, maintain stable climates, and produce the oxygen we breathe.

Today, the quality of the air we breathe and the water we drink does not augur well for our future health and well-being. Many people assume that the passage of the Clean Air Act in 1970 ushered in a new age. But the American Lung Association reports that 159 million Americans—55% of the population—are exposed to "unhealthy levels of air pollution." Meanwhile, the American Heart Association warns of a direct link between exposure to air pollution and heart disease and strokes. While it's true that U.S. waters are cleaner than they were three decades ago, data from the Environmental Protection Agency (EPA) shows that almost half of U.S. coastal waters fail to meet water-quality standards because they cannot support fishing or swimming. Each year, contaminated tap water makes as many as 7 million Americans sick. The chief cause is "non-point pollution," runoff that includes fertilizers and pesticides from farms and backyards as well as oil and chemical spills. On a global level, more than a billion people lack access to clean water; according to the United Nations, five times that number die each year from malaria and other illnesses associated with unsafe water.

Of all the Earth's critical environmental problems, one trumps the rest: climate change. Carol Browner, the EPA's chief from 1993 through 2001 (the longest term in the agency's history), calls climate change "the greatest environmental health problem the world has ever seen." Industry and people are spewing carbon dioxide from smokestacks and the tailpipes of their cars into the atmosphere, where a buildup of gases, acting like the glass in a greenhouse, traps the sun's heat. The 1990s was the warmest decade in more than a century, and 1998 saw the highest global temperatures ever. In an article about global climate change in the December 2003 issue of *Audubon*, David Malakoff wrote, "Among the possible consequences: rising sea levels that cause coastal communities to sink beneath the waves like a modern Atlantis, crop failures of biblical proportions, and once-rare killer storms that start to appear with alarming regularity."

Yet for all the doom and gloom, scientists and environmentalists hold out hope. When Russia recently ratified the Kyoto Protocol, it meant that virtually all of the world's industrialized nations—the United States, which has refused to sign, is a notable exception—have committed to cutting greenhouse gases. As Kyoto and other international agreements go into effect, a market is developing for cap-and-trade systems for carbon dioxide. In this country, two dozen big corporations, including British Petroleum, are cutting emissions. At least 28 American states have adopted their own policies. California, for example, has passed global warming legislation aimed at curbing emissions from new cars. Governor Arnold Schwarzenegger has also backed regulations requiring automakers to slash the amount of greenhouse gases they cause by up to 30% by 2016, setting a precedent for other states.

As Washington pushes a business-friendly agenda, states are filling in the policy vacuum in other areas, as well. California and New York are developing laws to preserve wetlands, which filter pollutants, prevent floods, and provide habitat for endangered wildlife.

By taking matters into their own hands, states and foreign countries will ultimately force Washington's. What industry especially abhors is a crazy quilt of varying rules. After all, it makes little sense for a company to invest a billion dollars in a power plant only to find out later that it has to spend even more to comply with another state's stricter emissions standards. Ford chairman and chief executive William Ford has lashed out at the states' "patchwork" approach because he and "other manufacturers will have a hard time responding." Further, he wrote in a letter to his company's top managers, "the prospect of 50 different requirements in 50 different states would be nothing short of chaos." The type of fears Ford expresses are precisely the reason federal laws protecting clean air and water came into being.

Governments must take the lead, but ecologically conscious consumers wield enormous influence, too. Over the past four decades, the annual use of pesticides has more than doubled, from 215 million pounds to 511 million pounds. Each year, these poisons cause $10 billion worth of damage to the environment and kill 72 million birds. The good news is that the demand for organic products is revolutionizing agriculture, in part by creating a market for natural alternatives for pest control. Some industry experts predict that by 2007 the organic industry will almost quadruple, to more than $30 billion.

E. O. Wilson touts "shade-grown" coffee as one of many "personal habitats that, if moderated only in this country, could contribute significantly to saving endangered species." In the mountains of Mexico and Central America, coffee grown beneath a dense forest canopy rather than in cleared fields helps provide refuge for dozens of wintering North American migratory bird species, from western tanagers to Baltimore orioles.

With conservation such a huge part of Americans' daily routine, recycling has become as ingrained a civic duty as obeying traffic lights. Californians, for their part, have cut their energy consumption by 10% each year since the state's 2001 energy crisis. "Poll after poll shows that about two-thirds of the American public—Democrat and Republican, urban and rural—consider environmental progress crucial," writes Carl Pope, director of the Sierra Club, in his recent book, *Strategic Ignorance.* "Clean air, clean water, wilderness preservation—these are such bedrock values that many polling respondents find it hard to believe that any politician would oppose them."

Terrorism and the economy clearly dwarfed all other issues in the 2004 presidential election. Even so, voters approved 120 out of 161 state and local conservation funding measures nationwide, worth a total of $3.25 billion. Anti-environment votes in the U.S. Congress and proposals floated by the like-minded Bush administration should not obscure the salient fact that so far there have been no changes to the major environmental laws. The potential for political fallout is too great.

The United States' legacy of preserving its natural heritage is the envy of the world. Our national park system alone draws more than 300 million visitors each year. Less well known is the 103-year-old national wildlife refuge system you'll learn about in this series. Its unique mission is to safeguard the nation's wild animals and plants on 540 refuges, protecting 700 species of birds and an equal number of other vertebrates; 282 of these species are either threatened or endangered. One of the many species particularly dependent on the invaluable habitat refuges afford is the bald eagle. Such safe havens, combined with the banning of the insecticide DDT and enforcement of the

Endangered Species Act, have led to the bald eagle's remarkable recovery, from a low of 500 breeding pairs in 1963 to 7,600 today. In fact, this bird, the national symbol of the United States, is about be removed from the endangered species list and downgraded to a less threatened status under the CITES, the Convention on International Trade in Endangered Species.

This vital treaty, upheld by the United States and 165 other participating nations (and detailed in this series), underscores the worldwide will to safeguard much of the Earth's magnificent wildlife. Since going into effect in 1975, CITES has helped enact plans to save tigers, chimpanzees, and African elephants. These species and many others continue to face dire threats from everything from poaching to deforestation. At the same time, political progress is still being made. Organizations like the World Wildlife Fund work tirelessly to save these species from extinction because so many millions of people care. China, for example, the most populous nation on Earth, is so concerned about its giant pandas that it has implemented an ambitious captive breeding program. That program's success, along with government measures prohibiting logging throughout the panda's range, may actually enable the remaining population of 1,600 pandas to hold its own—and perhaps grow. "For the People's Republic of China, pressure intensified as its internationally popular icon edged closer to extinction," wrote Gerry Ellis in a recent issue of *National Wildlife.* "The giant panda was not only a poster child for endangered species, it was a symbol of our willingness to ensure nature's place on Earth."

Whether people take a spiritual path to conservation or a pragmatic one, they ultimately arrive at the same destination. The sight of a bald eagle soaring across the horizon reassures us about nature's resilience, even as the clean air and water we both need to survive becomes less of a certainty. "The conservation of our natural resources and their proper use constitute the fundamental problem which underlies almost every other problem of our national life," President Theodore Roosevelt told Congress at the dawn of the conservation movement a century ago. His words ring truer today than ever.

Introduction: "Why Should We Care?"

Our nation's air and water are cleaner today than they were 30 years ago. After a century of filling and destroying over half of our wetlands, we now protect many of them. But the Earth is getting warmer, habitats are being lost to development and logging, and humans are using more water than ever before. Increased use of water can leave rivers, lakes, and wetlands without enough water to support the native plant and animal life. Such changes are causing plants and animals to go extinct at an increased rate. It is no longer a question of losing just the dodo birds or the passenger pigeons, argues David Quammen, author of *Song of the Dodo*: "Within a few decades, if present trends continue, we'll be losing *a lot* of everything."[1]

In the 1980s, E. O. Wilson, a Harvard biologist and Pulitzer Prize–winning author, helped bring the term *biodiversity* into public discussions about conservation. *Biodiversity*, short for "biological diversity," refers to the levels of organization for living things. Living organisms are divided and categorized into ecosystems (such as rain forests or oceans), by species (such as mountain gorillas), and by genetics (the genes responsible for inherited traits).

Wilson has predicted that if we continue to destroy habitats and pollute the Earth at the current rate, in 50 years, we could lose 30 to 50% of the planet's species to extinction. In his 1992 book, *The Diversity of Life*, Wilson asks: "Why should we care?"[2] His long list of answers to this question includes: the potential loss of vast amounts of scientific information that would enable the development of new crops, products, and medicines and the potential loss of the vast economic and environmental benefits of healthy ecosystems. He argues that since we have only a vague idea (even with our advanced scientific methods) of how ecosystems really work, it would be "reckless" to suppose that destroying species indefinitely will not threaten us all in ways we may not even understand.

THE BOOKS IN THE SERIES

In looking at environmental issues, it quickly becomes clear that, as naturalist John Muir once said, "When we try to pick out anything by itself, we find it hitched to everything else in the Universe."[3] For example, air pollution in one state or in one country can affect not only air quality in another place, but also land and water quality. Soil particles from degraded African lands can blow across the ocean and cause damage to far-off coral reefs.

The six books in this series address a variety of environmental issues: conservation, wildlife protection, water pollution, air quality, climate change, and environmental policy. None of these can be viewed as a separate issue. Air quality impacts climate change, wildlife, and water quality. Conservation initiatives directly affect water and air quality, climate change, and wildlife protection. Endangered species are touched by each of these issues. And finally, environmental policy issues serve as important tools in addressing all the other environmental problems that face us.

You can use the burning of coal as an example to look at how a single activity directly "hitches" to a variety of environmental issues. Humans have been burning coal as a fuel for hundreds of years. The mining of coal can leave the land stripped of vegetation, which erodes the soil. Soil erosion contributes to particulates in the air and water quality problems. Mining coal can also leave piles of acidic tailings that degrade habitats and pollute water. Burning any fossil fuel—coal, gas, or oil—releases large amounts of carbon dioxide into the atmosphere. Carbon dioxide is considered a major "greenhouse gas" that contributes to global warming—the gradual increase in the Earth's temperature over time. In addition, coal burning adds sulfur dioxide to the air, which contributes to the formation of acid rain—precipitation that is abnormally acidic. This acid rain can kill forests and leave lakes too acidic to support life. Technology continues to present ways to minimize the pollution that results from extracting and burning fossil fuels. Clean air and climate change policies guide states and industries toward implementing various strategies and technologies for a cleaner coal industry.

Each of the six books in this series—ENVIRONMENTAL ISSUES—introduces the significant points that relate to the specific topic and explains its relationship to other environmental concerns.

Book One: *Air Quality*

Problems of air pollution can be traced back to the time when humans first started to burn coal. *Air Quality* looks at today's challenges in fighting to keep our air clean and safe. The book includes discussions of air pollution sources—car and truck emissions, diesel engines, and many industries. It also discusses their effects on our health and the environment.

The Environmental Protection Agency (EPA) has reported that more than 150 million Americans live in areas that have unhealthy levels of some type of air pollution.[4] Today, more than 20 million Americans, over 6 million of whom are children, suffer from asthma believed to be triggered by pollutants in the air.[5]

In 1970, Congress passed the Clean Air Act, putting in place an ambitious set of regulations to address air pollution concerns. The EPA has identified and set standards for six common air pollutants: ground-level ozone, nitrogen oxides, particulate matter, sulfur dioxide, carbon monoxide, and lead.

The EPA has also been developing the Clean Air Rules of 2004, national standards aimed at improving the country's air quality by specifically addressing the many sources of contaminants. However, many conservation organizations and even some states have concerns over what appears to be an attempt to weaken different sections of the 1990 version of the Clean Air Act. The government's environmental protection efforts take on increasing importance because air pollution degrades land and water, contributes to global warming, and affects the health of plants and animals, including humans.

Book Two: *Climate Change*

Part of science is observing patterns, and scientists have observed a global rise in temperature. *Climate Change* discusses the sources and effects of global warming. Scientists attribute this accelerated change to human activities such as the burning of fossil fuels that emit greenhouse gases (GHG).[6] Since the 1700s, we have been cutting down the trees that help remove carbon dioxide from the atmosphere, and have increased the amount of coal, gas, and oil we burn, all of which add carbon dioxide to the atmosphere. Science tells us that these human activities have caused greenhouse gases—carbon dioxide (CO_2), methane (CH_4), nitrous oxide (N_2O), hydrofluorocarbons (HFCs), perfluorocarbons (PFCs), and sulfur hexafluoride (SF_6)—to accumulate in the atmosphere.[7]

If the warming patterns continue, scientists warn of more negative environmental changes. The effects of climate change, or global warming, can be seen all over the world. Thousands of scientists are predicting rising sea levels, disturbances in patterns of rainfall and regional weather, and changes in ranges and reproductive cycles of plants and animals. Climate change is already having some effects on certain plant and animal species.[8]

Many countries and some American states are already working together and with industries to reduce the emissions of greenhouse gases. Climate change is an issue that clearly fits noted scientist Rene Dubois's advice: "Think globally, act locally."

Book Three: *Conservation*

Conservation considers the issues that affect our world's vast array of living creatures and the land, water, and air they need to survive.

One of the first people in the United States to put the political spotlight on conservation ideas was President Theodore Roosevelt. In the early 1900s, he formulated policies and created programs that addressed his belief that: "The nation behaves well if it treats the natural resources as assets which it must turn over to the next generation increased, and not impaired, in value."[9] In the 1960s, biologist Rachel Carson's book, *Silent Spring*, brought conservation issues into the public eye. People began to see that polluted land, water, and air affected their health. The 1970s brought the creation of the United States Environmental Protection Agency (EPA) and passage of many federal and state rules and regulations to protect the quality of our environment and our health.

Some 80 years after Theodore Roosevelt established the first National Wildlife Refuge in 1903, Harvard biologist

E. O. Wilson brought public awareness of conservation issues to a new level. He warned:

> . . . the worst thing that will probably happen—in fact is already well underway—is not energy depletion, economic collapse, conventional war, or even the expansion of totalitarian governments. As terrible as these catastrophes would be for us, they can be repaired within a few generations. The one process now ongoing that will take million of years to correct is the loss of genetic species diversity by the destruction of natural habitats. This is the folly our descendants are least likely to forgive us.[10]

To heed Wilson's warning means we must strive to protect species-rich habitats, or "hotspots," such as tropical rain forests and coral reefs. It means dealing with conservation concerns like soil erosion and pollution of fresh water and of the oceans. It means protecting sea and land habitats from the over-exploitation of resources. And it means getting people involved on all levels—from national and international government agencies, to private conservation organizations, to the individual person who recycles or volunteers to listen for the sounds of frogs in the spring.

Book Four: *Environmental Policy*

One approach to solving environmental problems is to develop regulations and standards of safety. Just as there are rules for living in a community or for driving on a road, there are environmental regulations and policies that work toward protecting our health and our lands. *Environmental Policy* discusses the regulations and programs that have been crafted to address environmental issues at all levels—global, national, state, and local.

Today, as our resources become increasingly limited, we witness heated debates about how to use our public lands and how to protect the quality of our air and water. Should we allow drilling in the Arctic National Wildlife Refuge? Should we protect more marine areas? Should we more closely regulate the emissions of vehicles, ships, and industries? These policy issues, and many more, continue to make news on a daily basis.

In addition, environmental policy has taken a place on the international front. Hundreds of countries are working together in a variety of ways to address such issues as global warming, air pollution, water pollution and supply, land preservation, and the protection of endangered species. One question the United States continues to debate is whether to sign the 1997 Kyoto Protocol, the international agreement designed to decrease the emissions of greenhouse gases.

Many of the policy tools for protecting our environment are already in place. It remains a question how they will be used—and whether they will be put into action in time to save our natural resources and ourselves.

Book Five: *Water Pollution*

Pollution can affect water everywhere. Pollution in lakes and rivers is easily seen. But water that is out of our plain view can also be polluted with substances such as toxic chemicals, fertilizers, pesticides, oils, and gasoline. *Water Pollution* considers issues of concern to our surface waters, our groundwater, and our oceans.

In the early 1970s, about three-quarters of the water in the United States was considered unsafe for swimming and fishing. When Lake Erie was declared "dead" from pollution and a river feeding it actually caught on fire, people decided that the national government had to take a stronger role in protecting our resources. In 1972, Congress passed the Clean Water Act, a law whose objective "is to restore and maintain the chemical, physical, and biological integrity of the Nation's waters."[11] Today, over 30 years later, many lakes and rivers have been restored to health. Still, an estimated 40% of our waters are still unsafe to swim in or fish.

Less than 1% of the available water on the planet is fresh water. As the world's population grows, our demand for drinking and irrigation water increases. Therefore, the quantity of available water has become a major global issue. As Sandra Postel, a leading authority on international freshwater issues, says, "Water scarcity is now the single biggest threat to global food production."[12] Because there are many competing demands for water, including the needs of habitats, water pollution continues to become an even more serious problem each year.

Book Six: *Wildlife Protection*

For many years, the word *wildlife* meant only the animals that people hunted for food or for sport. It was not until 1986 that the Oxford English Dictionary defined *wildlife* as "the native fauna and flora of a particular region."[13] *Wildlife Protection* looks at overexploitation—for example, overfishing or collecting plants and animals for illegal trade—and habitat loss. Habitat loss can be the result of development, logging, pollution, water diverted for human use, air pollution, and climate change.

Also discussed are various approaches to wildlife protection. Since protection of wildlife is an issue of global concern, it is addressed here on international as well as on national and local levels. Topics include voluntary international organizations such as the International Whaling Commission and the CITES agreements on trade in endangered species. In the United States, the Endangered Species Act provides legal protection for more than 1,200 different plant and animal species. Another approach to wildlife protection includes developing partnerships among conservation organizations, governments, and local people to foster economic incentives to protect wildlife.

CONSERVATION IN THE UNITED STATES

Those who first lived on this land, the Native American peoples, believed in general that land was held in common, not to be individually owned, fenced, or tamed. The white settlers from Europe had very different views of land. Some believed the New World was a Garden of Eden. It was a land of opportunity for them, but it was also a land to be controlled and subdued. Ideas on how to treat the land often followed those of European thinkers like John Locke, who believed that "Land that is left wholly to nature is called, as indeed it is, waste." [14]

The 1800s brought another way of approaching the land. Thinkers such as Ralph Waldo Emerson, John Muir, and Henry David Thoreau celebrated our human connection with nature. By the end of the 1800s, some scientists and policymakers were noticing the damage humans have caused to the land. Leading public officials preached stewardship and wise use of our country's resources. In 1873, Yellowstone National Park was set up. In 1903, the first National Wildlife Refuge was established.

However, most of the government practices until the middle of the 20th century favored unregulated development and use of the land's resources. Forests were clear cut, rivers were dammed, wetlands were filled to create farmland, and factories were allowed to dump their untreated waste into rivers and lakes.

In 1949, a forester and ecologist named Aldo Leopold revived the concept of preserving land for its own sake. But there was now a biological, or scientific, reason for conservation, not just a spiritual one. Leopold declared: "All ethics rest upon a single premise: that the individual is a member of a community of interdependent parts. . . . A thing is right when it tends to preserve the integrity and stability and beauty of the biotic community. It is wrong when it tends otherwise." [15]

The fiery vision of these conservationists helped shape a more far-reaching movement that began in the 1960s. Many credit Rachel Carson's eloquent and accessible writings, such as her 1962 book *Silent Spring*, with bringing environmental issues into people's everyday language. When the Cuyahoga River in Ohio caught fire in 1969 because it was so polluted, it captured the public attention. Conservation was no longer just about protecting land that many people would never even see, it was about protecting human health. The condition of the environment had become personal.

In response to the public outcry about water and air pollution, the 1970s saw the establishment of the EPA. Important legislation to protect the air and water was passed. National standards for a cleaner environment were set and programs were established to help achieve the ambitious goals. Conservation organizations grew from what had started as exclusive white men's hunting clubs to interest groups with a broad membership base. People came together to demand changes that would afford more protection to the environment and to their health.

Since the 1960s, some presidential administrations have sought to strengthen environmental protection and to protect more land and national treasures. For example, in 1980, President Jimmy Carter signed an act that doubled the amount of protected land in Alaska and renamed it the Arctic National Wildlife Refuge. Other administrations, like those of President Ronald Reagan, sought to dismantle many earlier environmental protection initiatives.

The environmental movement, or environmentalism, is not one single, homogeneous cause. The agencies, individuals, and organizations that work toward protecting the environment vary as widely as the habitats and places they seek to protect. There are individuals who begin grass-roots efforts—people like Lois Marie Gibbs, a former resident of the polluted area of Love Canal, New York, who founded the Center for Health, Environment and Justice. There are conservation organizations, like The Nature Conservancy, the World Wildlife Fund (WWF), and Conservation International, that sponsor programs to preserve and protect habitats. There are groups that specialize in monitoring public policy and legislation—for example, the Natural Resources Defense Council and Environmental Defense. In addition, there are organizations like the Audubon Society and the National Wildlife Federation whose focus is on public education about environmental issues. Perhaps from this diversity, just like there exists in a healthy ecosystem, will come the strength and vision environmentalism needs to deal with the continuing issues of the 21st century.

INTERNATIONAL CONSERVATION EFFORTS

In his book *Biodiversity*, E. O. Wilson cautions that biological diversity must be taken seriously as a global resource for three reasons. First, human population growth is accelerating the degrading of the environment, especially in tropical countries. Second, science continues to discover new uses for biological diversity—uses that can benefit human health and protect the environment. And third, much biodiversity is being lost through extinction, much of it in the tropics. As Wilson states, "We must hurry to acquire the knowledge on which a wise policy of conservation and development can be based for centuries to come." [16]

People organize themselves within boundaries and borders. But oceans, rivers, air, and wildlife do not follow such rules. Pollution or overfishing in one part of an ocean can easily degrade the quality of another country's resources. If one country diverts a river, it can destroy another country's wetlands or water resources. When Wilson cautions us that we must hurry to develop a wise conservation policy, he means a policy that will protect resources all over the world.

To accomplish this will require countries to work together on critical global issues: preserving biodiversity, reducing global warming, decreasing air pollution, and protecting the oceans. There are many important international efforts already going on to protect the resources of our planet. Some efforts are regulatory, while others are being pursued by nongovernmental organizations or private conservation groups.

Countries volunteering to cooperate to protect resources is not a new idea. In 1946, a group of countries established the International Whaling Commission (IWC). They recognized that unregulated whaling around the world had led to severe declines in the world's whale populations. In 1986, the IWC declared a moratorium on whaling, which is still in effect, until the populations have recovered. [17] Another example of international cooperation occurred in 1987 when various countries signed the Montreal Protocol to reduce the emissions of ozone-depleting gases. It has been a huge success, and perhaps has served as a model for other international efforts, like the 1997 Kyoto Protocol, to limit emissions of greenhouse gases.

Yet another example of international environmental cooperation is the CITES agreement (the Convention on International Trade in Endangered Species of Wild Fauna and Flora), a legally binding agreement to ensure that the international trade of plants and animals does not threaten the species' survival. CITES went into force in 1975 after 80 countries agreed to the terms. Today, it has grown to include more than 160 countries. This make CITES among the largest conservation agreements in existence.[18]

Another show of international conservation efforts are governments developing economic incentives for local conservation. For example, in 1996, the International Monetary Fund (IMF) and the World Wildlife Fund (WWF) established a program to relieve poor countries of debt. More than 40 countries have benefited by agreeing to direct some of their savings toward environmental programs in the "Debt-for-Nature" swap programs.[19]

It is worth our time to consider the thoughts of two American conservationists and what role we, as individuals, can play in conserving and protecting our world. E. O. Wilson has told us that "Biological Diversity—'biodiversity' in the new parlance—is the key to the maintenance of the world as we know it."[20] Aldo Leopold, the forester who gave Americans the idea of creating a "land ethic," wrote in 1949 that: "Having to squeeze the last drop of utility out of the land has the same desperate finality as having to chop up the furniture to keep warm."[21] All of us have the ability to take part in the struggle to protect our environment and to save our endangered Earth.

ENDNOTES

1   Quammen, David. *Song of the Dodo*. New York: Scribner, 1996, p. 607.

2   Wilson, E. O. *Diversity of Life*. Cambridge, MA: Harvard University Press, 1992, p. 346.

3   Muir, John. *My First Summer in the Sierra*. San Francisco: Sierra Club Books, 1988, p. 110.

4 Press Release. *EPA Newsroom: EPA Issues Designations on Ozone Health Standards.* April 15, 2004. Available online at *http://www.epa.gov/newsroom/.*

5 The Environmental Protection Agency. EPA Newsroom. *May is Allergy Awareness Month.* May 2004. Available online at *http://www.epa.gov/newsroom/allergy_month.htm.*

6 Intergovernmental Panel on Climate Change (IPCC). Third Annual Report, 2001.

7 Turco, Richard P. *Earth Under Siege: From Air Pollution to Global Change.* New York: Oxford University Press, 2002, p. 387.

8 Intergovernmental Panel on Climate Change. *Technical Report V: Climate Change and Biodiversity.* 2002. Full report available online at *http://www.ipcc.ch/pub/tpbiodiv.pdf.*

9 "Roosevelt Quotes." American Museum of Natural History. Available online at *http://www.amnh.org/common/faq/quotes.html.*

10 Wilson, E. O. *Biophilia.* Cambridge, MA: Harvard University Press, 1986, pp. 10–11.

11 Federal Water Pollution Control Act. As amended November 27, 2002. Section 101 (a).

12 Postel, Sandra. *Pillars of Sand.* New York: W. W. Norton & Company, Inc., 1999. p. 6.

13 Hunter, Malcolm L. *Wildlife, Forests, and Forestry: Principles of Managing Forest for Biological Diversity.* Englewood Cliffs, NJ: Prentice-Hall, 1990, p. 4.

14 Dowie, Mark. *Losing Ground: American Environmentalism at the Close of the Twentieth Century.* Cambridge, MA: MIT Press, 1995, p. 113.

15 Leopold, Aldo. *A Sand County Almanac.* New York: Oxford University Press, 1949.

16 Wilson, E. O., ed. *Biodiversity.* Washington, D.C.: National Academies Press, 1988, p. 3.

17 International Whaling Commission Information 2004. Available online at *http://www.iwcoffice.org/commission/iwcmain.htm.*

18 *Discover CITES: What is CITES?* Fact sheet 2004. Available online at *http://www.cites.org/eng/disc/what.shtml.*

19 *Madagascar's Experience with Swapping Debt for the Environment.* World Wildlife Fund Report, 2003. Available online at *http://www.conservationfinance.org/WPC/WPC_documents/ Apps_11_Moye_Paddack_v2.pdf.*

20 Wilson, *Diversity of Life,* p. 15.

21 Leopold.

Air Quality Issues and Challenges

How Clean Is Our Air?

In 1970, the Clean Air Act was passed. Because of this important legislation, the air quality in the United States has improved substantially over the last 35 years. But as the following 2004 American Lung Association (ALA) report explains, in many places, the air is still not clean enough. The Environmental Protection Agency (EPA) has reported that more than 150 million Americans live in areas with unhealthy levels of some type of air pollution.

The EPA report identifies particulate matter (particle pollution) and ozone pollution as two areas of concern. "Particulate matter" refers to all the tiny pieces of smoke, dust, and dirt found in the air. Sometimes this kind of pollution appears as a mucky brown haze. Cars, trucks, fires, and industries all contribute particulate matter to the air.

The ozone that causes health problems is created through a chemical reaction, occurring in sunlight, between nitrogen oxides (NOx) and volatile organic compounds (VOCs). It is the same ozone found 10 to 20 miles (16 to 32 km) above the Earth that protects us from the sun's radiation. But when ozone gas is found too close to the ground, it presents a health hazard. Cars, trucks, diesel engines, and industries all emit NOx and VOCs that react in sunlight to form ozone.

A section of this ALA report introduces measures that have been taken to decrease the pollution emissions from a variety of sources. The review highlights the American democratic process in action. Over the past few years, a group of environmental organizations and the American Lung Association have taken legal action against the EPA for what they have felt was the current government administration's attempt to weaken the Clean Air Act. The result of the legal settlement was that the EPA agreed to review and revise its regulations as appropriate. Since that time, the EPA has passed stronger rules on regulating diesel engines.

—The Editor

State of the Air 2004
from the American Lung Association

EXECUTIVE SUMMARY

Millions of Americans were subjected to dangerous levels of air pollution during the years 2000 to 2002. The *American Lung Association State of the Air: 2004* presents information on air pollution on a state-by-state, county-by-county basis, using the most up-to-date quality assured data available for nation-wide comparisons.

For the first time, in addition to its traditional focus on ozone pollution, the *American Lung Association State of the Air: 2004* expands to include a county-level report card on particle pollution, a pollutant that represents risks to the lives of far too many Americans. In addition, this year's report shows that ozone remains a persistent threat across large parts of the United States.

Some of the facts from this report card on air pollution are below, taking a look at the nation as a whole:

Nearly half the U.S. population—47%—lives in areas with unhealthful levels of ozone.

Counties that were graded F for ozone levels have a combined population of 136 million. Almost half of America is living in counties where the air quality places them at risk for decreased lung function, respiratory infection, lung inflammation and aggravation of respiratory illness.

Over one quarter—28%—of the U.S. population lives in areas with unhealthful short-term levels of particle pollution.

Over 81 million Americans live in areas where they are exposed to unhealthful short-term levels of particle pollution. Short-term, or acute, exposure to particle pollution has been shown to increase heart attacks, strokes, and emergency-room visits for respiratory ailments and cardiovascular disease, and most importantly, increase the risk of death.

Nearly one quarter—23%—of the U.S. population lives in areas with unhealthful year-round levels of particle pollution.

Sixty-six million Americans suffer from chronic exposure to particle pollution. Even when levels are fairly low, over time exposure to particles can increase risk of hospitalization for asthma, damage to the lungs and significantly increase the risk of premature death.

Over half—55%—of the U.S. population lives in counties which have unhealthy levels of either ozone or particle pollution.

Approximately 159 million Americans live in 441 counties where they are exposed to unhealthy levels of air pollution in the form of either ozone or short-term or year-round levels of particles.

About 46 million Americans—nearly 16%—live in 48 counties with unhealthy levels of all three: ozone and short-term and year-round particle pollution.

With the risks from airborne pollution so great, the American Lung Association seeks to inform people who may be in danger. Many groups are at greater risk because of their age or the presence of a chronic lung or cardiovascular disease. Those groups include:

Adult and Pediatric Asthma

Nearly 7.5 million adults and nearly 3 million children with asthma live in parts of the United States with very high levels of ozone. Nearly 4.5 million adults and 1.8 million children with asthma live in areas with high levels of short-term particle pollution. Three and a half million adults and nearly one and a half million children with asthma live in counties with unhealthful levels of year-round particle pollution.

Older and Younger

Over 15 million adults 65 and over and 29 million children age 14 and under live in counties with unhealthful ozone levels. Over 9.3 million seniors and over 17.8 million children live in counties, that have unhealthful short-term levels of particle pollution. Close to 7.6 million seniors and over 14 million children live in counties with unhealthful levels of year-round particle pollution.

Chronic Bronchitis and Emphysema

Over 4.4 million people with chronic bronchitis and 1.5 million with emphysema live in counties with unhealthful ozone levels. Some 2.6 million people with chronic bronchitis and 888,000 with emphysema live in counties with unhealthful levels of short-term particle pollution. Over 2 million people with chronic bronchitis and nearly three quarters of a million (720,000) with emphysema live in counties with unhealthful year-round levels of particle pollution.

Cardiovascular Disease

Over 16.7 million Americans with cardiovascular diseases live in areas with unhealthful levels of short-term particle pollution; 13.6 million live in counties with unhealthful levels of year-round particle pollution. Cardiovascular diseases include heart disease, heart attacks and strokes.

In addition to providing specific grades for each county with ozone and particle pollution monitor, the *American Lung Association State of the Air: 2004* also discusses key steps needed to improve the air we all breathe. Those steps include:

Protect the Clean Air Act

The American Lung Association is greatly concerned about threats to one of the most effective public health laws ever passed, the Clean Air Act. Threats come from two areas: legislative and regulatory proposals to roll back key provisions of the law, and continued delays in putting into place what the science tells us is needed to clean up air pollution. The American Lung Association has taken legal action to protect this valuable clean air tool, and encourages everyone to tell his or her members of Congress to protect the Clean Air Act.

Clean up Dirty Power Plants

Old coal-fired power plants have become some of the biggest industrial contributors to our unhealthy air, especially to the level of particle pollution in the eastern United States. The toll of death, disease and environmental destruction caused by coal-fired power plant pollution continues to mount. The Environmental Protection Agency (EPA) issued proposed rules in 2003 that would give states the tools to clean up these plants. The rules need to be stronger and, most of all, made final so work can begin.

Clean up Dirty Diesel

While new rules to regulate emissions of diesel truck and buses will make a great deal of difference in the quality of our air, these rules alone will not be enough. EPA also must take steps to control heavy equipment and other nonroad diesel engines and fuel to the same degree as diesel buses and trucks. In fact, heavy equipment diesel engines (such as bulldozers, excavators, tractors, electric generators and forklifts) are a larger source of emissions than diesel trucks and buses.

Individuals can do a great deal to help reduce air pollution outdoors as well. Here are some simple, but effective ways:

Reduce Driving

Combine trips, walk, bike, carpool or vanpool and use buses, subways or other alternatives to driving. Vehicle emissions are a major source of air pollution. Support community plans that provide ways to get around that don't require a car, such as more sidewalks, bike trails and transit systems.

Fill up Cars After Dark

Gasoline emissions evaporating while you fill up your gas tank contribute to forming ozone. Filling up after dark helps prevents the sun from turning those gases into ozone.

Don't Burn Wood or Trash

Burning firewood and trash are some of the largest sources of particles in many parts of the country. Convert your woodstoves into natural gas, which has far fewer emissions. Dispose of trash properly.

Get involved in your community's review of the air pollution plans and support state and local efforts to clean up air pollution.

1. INTRODUCTION

Each year, the American Lung Association assesses the toll that air pollution places on our nation's ability to breathe. This year's look at county-level air quality expands by more than two times the information provided in previous reports. The *American Lung Association State of the Air: 2004* for the first time examines an additional pollutant, PM2.5 or particle pollution, in two new measures: the short-term exposure, which are occasional spikes in particle pollution from relatively infrequent events (although these spikes may last hours to days); and the year-round or chronic exposure from particles produced routinely in the environment. In addition, the report examines the latest quality-assured data on ozone for each county that has an ozone monitor, as it has for five years.

PARTICLE POLLUTION

Particle pollution has emerged as a widespread problem, especially in large parts of the eastern United States and California. This report looks first at the presence of particle pollution by U.S. Environmental Protection Agency (EPA) region. . . . This report also includes tables with each state's short-term and year-round particle grades for each county with a particle monitor. These data come from a network of monitors in over 700 counties established in 1998 and 1999 following EPA's adoption of a new health standard to address particle pollution in 1997. This is the first such analysis of the three years of complete data from those monitors. [You can read the full report online at *http://lungaction.org/reports/sota04intro.html.*]

OZONE POLLUTION

Ozone continues to be the most pervasive air pollutant, and remains a present danger despite decreases in levels of this pollutant across the nation since 1980. During the 1990s, ozone concentrations remained remarkably and uncomfortably unchanged. EPA's own records show this stagnation. However, EPA's data are now showing a slight trend toward lower ozone readings, a trend also reflected by the analysis in this report. This slight decline also comes in the face of a particularly hot summer in 2002 when many cities reported "Code Red" days, when air pollution levels reached unhealthful levels for all populations. EPA speculates that these declines may be coming from controls put in place to clean up coalfired power plants in the eastern United States. If so, this trend will likely persist in future reports, as work is expected to continue in this period as additional control measures are installed on plants through May 2004.

MILLIONS ARE AT RISK

For the first time, the data allow a tally of the number of people who live in counties where monitors show they have unhealthy levels of air pollution, in the form of either ozone or short-term or year-round levels of particle pollution.

- 159 million Americans—55% of the U.S. population—live in 441 counties where they are exposed to unhealthy levels of air pollution in the form of either ozone or short-term levels or year-round levels of particle pollution.

- 46 million Americans—nearly 16% of the population—live in 48 counties with unhealthy levels of all three: ozone and particle pollution in both short-term and year-round levels.

OZONE

Even with the slight downturn in ozone levels, this report finds that nearly half of the people in the United States—47%—live in counties with unhealthful levels of ozone pollution. Included are nearly 136 million Americans, an estimate that understates the problem considerably since it only includes counties where ozone monitors exist and have accumulated three years of data. Of those 136 million, many of those are especially at risk.

PARTICLE POLLUTION

All too many who live in areas with unhealthful ozone levels also face a second, even more dangerous threat: particle pollution. This report estimates that millions live in areas with unhealthful either short-term or year-round levels of particle pollution:

- 81 million live in counties with unhealthful short-term levels of particle pollution and

- 66 million live in counties with chronically unhealthful particle levels.

Those who are particularly vulnerable to ozone are also at greater risk from particles. Unfortunately, particle pollution also threatens another large group: people with cardiovascular diseases. All totaled, millions of especially endangered Americans are living in areas where particle pollution levels place them at risk.

THE BASIS FOR THE AMERICAN LUNG ASSOCIATION *STATE OF THE AIR* REPORT

Because millions are exposed and millions are at risk, the American Lung Association produces the *American Lung Association State of the Air* each year to alert individuals, families, industry and government leaders to the dangers inherent in the air we breathe.

In 2000, the American Lung Association initiated its *State of the Air* annual assessment to provide citizens with easy-to-understand air pollution summaries of the quality of the air in their communities that are based on concrete data and sound science. Counties are assigned grades ranging from "A" through "F" based on how often their air quality crosses into the "unhealthful" categories of EPA's Air Quality Index for ground-level ozone (smog) pollution, and now, for short-term particle pollution.

The Air Quality Index is, in turn, based on the national air quality standards. The air quality standard for ozone used as the basis for this report, 0.08 parts per million averaged over an eight-hour period, was adopted by the EPA in 1997 based on the most recent health effects information. For particle pollution, the Air Quality Index is based on, but is more conservative than the PM2.5 24-hour national standard. Also adopted in 1997, the national standard for PM2.5 24-hour levels is 65 $\mu g/m^3$. However, EPA set the Air Quality Index for particles to acknowledge that levels below 65 $\mu g/m^3$ are harmful to public health.

To evaluate the year-round levels of particle pollution for any monitored county, the *American Lung Association State of the Air: 2004* uses the decision of EPA in its determination whether the county met or failed to meet the national air quality standards. . . .

The grades in this report are assigned based on the quality of the air in areas, and do not reflect an assessment of efforts to implement controls that improve air quality. The grades should not be interpreted as an evaluation of the work of any state or local air pollution control programs.

What Pollutes the Air We Breathe?

For centuries, the skies have been darkened by burning coal. In the mid-1800s, cities in Europe and the United States experienced periods of intense air pollution from the coal-fired industries of the Industrial Revolution. The term *smog* was first used to refer to a mixture of fog and smoke. Today, the pollutants in city smog are derived from automobile exhaust and industrial emissions.[1]

In 1970, the U.S. Environmental Protection Agency (EPA) revised the Clean Air Act of 1963, creating an ambitious set of regulations to address the problem of dirty air. The EPA works with the states to develop plans and to enforce the regulations to decrease the emissions that cause pollution.

The following collection of factsheets from the EPA offers some basic information on air pollution. In order to understand the health and policy issues, it is important to know the common pollutants, their sources, and their risks to human health and the environment.

The EPA has identified six common air pollutants for which standards have been set. These include ground-level ozone, nitrogen oxides, particulate matter, sulfur dioxide, carbon monoxide, and lead. Over the last 20 years, the levels of all of these pollutants have decreased. The EPA also works with states to reduce the release of another 188 toxic pollutants into the air. Examples of these include benzene, dioxin, mercury, and cadmium.

In April 2004, the EPA announced that 474 counties across the United States, with a combined population of more than 150 million people, were out of compliance with the federal health-based smog standards.[2] As part of a legal settlement with Environmental Defense and Earthjustice, the EPA agreed to release a final plan in March 2005 to reduce emissions from power plants and industrial sources.[3]

—The Editor

1. Turco, Richard P. *Earth Under Siege: From Air Pollution to Global Change* New York: Oxford University Press, 2002.
2. Press Release. *EPA Newsroom: EPA Issues Designations on Ozone Health Standards.* April 15, 2004. Available online at *http://www.epa.gov/newsroom/*.
3. Environmental Defense News Release. *EPA Clean Air Plan Must Ensure No Child Left Behind*. April 15, 2004.

Air Pollutant Factsheets
from the U.S. Environmental Protection Agency

WHAT ARE THE SIX COMMON AIR POLLUTANTS?

EPA has set national air quality standards for six common pollutants (also referred to as "criteria" pollutants):

- Ozone

- Particulate Matter

- Carbon Monoxide

- Lead

- Nitrogen Dioxide

- Sulfur Dioxide.

The Clean Air Act established two types of National Ambient Air Quality Standards.

"Primary" standards are designed to establish limits to protect public health, including the health of "sensitive" populations such as asthmatics, children, and the elderly.

"Secondary" standards set limits to protect public welfare, including protection against decreased visibility and damage to animals, crops, vegetation, and buildings.

For each of these pollutants, EPA tracks two kinds of air pollution trends: air concentrations based on actual measurements of pollutant concentrations in the ambient (outside) air at selected monitoring sites throughout the country, and emissions based on engineering estimates of the total tons of pollutants released into the air each year. Despite the progress made in the last 30 years, millions of people live in counties with monitor data showing unhealthy air for one or more of the six common pollutants.

GROUND-LEVEL OZONE: WHAT IS IT? WHERE DOES IT COME FROM?

Ozone (O_3) is a gas composed of three oxygen atoms. It is not usually emitted directly into the air, but at ground level is created by a chemical reaction between oxides of nitrogen (NOx) and volatile organic compounds (VOC) in the presence of heat and sunlight. Ozone has the same chemical structure whether it occurs miles above the earth or at ground level and can be "good" or "bad," depending on its location in the atmosphere. "Good" ozone occurs naturally in the stratosphere approximately 10 to 30 miles (16 to 48 km) above the earth's surface and forms a layer that protects life on earth from the sun's harmful rays. In the earth's lower atmosphere, ground-level ozone is considered "bad."

$$VOC + NOx + Heat + Sunlight = Ozone$$

Motor vehicle exhaust and industrial emissions, gasoline vapors, and chemical solvents are some of the major sources of NOx and VOC, that help to form ozone. Sunlight and hot weather cause ground-level ozone to form in harmful concentrations in the air. As a result, it is known as a summertime air pollutant. Many urban areas tend to have high levels of "bad" ozone, but even rural areas are also subject to increased ozone levels because wind carries ozone and pollutants that form it hundreds of miles away from their original sources.

Chief Causes for Concern
Ground-level Ozone

- Triggers a variety of health problems even at very low levels

- May cause permanent lung damage after long-term exposure

- Damages plants and ecosystems.

The Summertime Pollutant

Peak ozone levels typically occur during hot, dry, stagnant summertime conditions. The length of the ozone season varies from one area of the United States to another. Southern and Southwestern states may have an ozone season that lasts nearly the entire year.

Ozone Can Be Transported Over Long Distances

Ozone and the chemicals that react to form it can be carried hundreds of miles from their origins, causing air pollution over wide regions. Millions of Americans live in areas where ozone levels exceed EPA's health-based air quality standards, primarily in parts of the Northeast, the Lake Michigan area, parts of the Southeast, southeastern Texas, and parts of California.

Health and Environmental Impacts of Ground-level Ozone

Ground-level ozone even at low levels can adversely affect everyone. It can also have detrimental effects on plants and ecosystems.

Health Problems

- Ozone can irritate lung airways and cause inflammation much like a sunburn. Other symptoms include wheezing, coughing, pain when taking a deep breath, and breathing difficulties during exercise or outdoor activities. People with respiratory problems are most vulnerable, but even healthy people that are active outdoors can be affected when ozone levels are high.

- Repeated exposure to ozone pollution for several months may cause permanent lung damage. Anyone who spends time outdoors in the summer is at risk, particularly children and other people who are active outdoors.

- Even at very low levels, ground-level ozone triggers a variety of health problems including aggravated asthma, reduced lung capacity, and increased susceptibility to respiratory illnesses like pneumonia and bronchitis.

Plant and Ecosystem Damage

- Ground-level ozone interferes with the ability of plants to produce and store food, which makes them more susceptible to disease, insects, other pollutants, and harsh weather.

- Ozone damages the leaves of trees and other plants, ruining the appearance of cities, national parks, and recreation areas.

- Ozone reduces crop and forest yields and increases plant vulnerability to disease, pests, and harsh weather.

NOx: WHAT IS IT? WHERE DOES IT COME FROM?

Nitrogen oxides, or NOx, is the generic term for a group of highly reactive gases, all of which contain nitrogen and oxygen in varying amounts. Many of the nitrogen oxides are colorless and odorless. However, one common pollutant, nitrogen dioxide (NO_2) along with particles in the air can often be seen as a reddish-brown layer over many urban areas.

Nitrogen oxides form when fuel is burned at high temperatures, as in a combustion process. The primary sources of NOx are motor vehicles, electric utilities, and other industrial, commercial, and residential sources that burn fuels.

Chief Causes for Concern
NOx

- is one of the main ingredients involved in the formation of ground-level ozone, which can trigger serious respiratory problems.

- reacts to form nitrate particles, acid aerosols, as well as NO_2, which also cause respiratory problems.

- contributes to formation of acid rain.

- contributes to nutrient overload that deteriorates water quality.

- contributes to atmospheric particles, that cause visibility impairment most noticeable in national parks.

- reacts to form toxic chemicals.

- contributes to global warming.

NOx and the pollutants formed from NOx can be transported over long distances, following the pattern of prevailing winds in the U.S. This means that problems associated with NOx are not confined to areas where NOx are emitted. Therefore, controlling NOx is often most effective if done from a regional perspective, rather than focusing on sources in one local area.

NOx Emissions Are Increasing

Since 1970, EPA has tracked emissions of the six principal air pollutants—carbon monoxide, lead, nitrogen oxides, particulate matter, sulfur dioxide, and volatile organic compounds. Emissions of all of these pollutants have decreased significantly except for NOx which has increased approximately 10 percent over this period.

Health and Environmental Impacts of NOx

NOx causes a wide variety of health and environmental impacts because of various compounds and derivatives in the family of nitrogen oxides, including nitrogen dioxide, nitric acid, nitrous oxide, nitrates, and nitric oxide.

Ground-level Ozone (**Smog**) is formed when NOx and volatile organic compounds (VOCs) react in the presence of heat and sunlight. Children, people with lung diseases such as asthma, and people who work or exercise outside are suscep- *sensitive* tible to adverse effects such as damage to lung tissue and reduction in lung function. Ozone can be transported by wind currents and cause health impacts far from original sources. Millions of Americans live in areas that do not meet the health standards for ozone. Other impacts from ozone include damaged vegetation and reduced crop yields.

Acid Rain—NOx and sulfur dioxide react with other substances in the air to form acids which fall to earth as rain, fog, snow or dry particles. Some may be carried by wind for hundreds of miles. Acid rain causes deterioration of cars, buildings and historical monuments; and causes lakes and streams to become acidic and unsuitable for many fish.

Particles—NOx reacts with ammonia, moisture, and other compounds to form nitric acid and related particles. Human health concerns include effects on breathing and the respiratory system, damage to lung tissue, and premature death. Small particles penetrate deeply into sensitive parts of the lungs and can cause or worsen respiratory disease such as emphysema and bronchitis, and aggravate existing heart disease.

Water Quality Deterioration—Increased nitrogen loading in water bodies, particularly coastal estuaries, upsets the chemical balance of nutrients used by aquatic plants and animals. Additional nitrogen accelerates "eutrophication," which leads to oxygen depletion and reduces fish and shellfish populations. NOx emissions in the air are one of the largest sources of nitrogen pollution in the Chesapeake Bay.

Global Warming—One member of the NOx, nitrous oxide, is a greenhouse gas. It accumulates in the atmosphere with other greenhouse gasses causing a gradual rise in the earth's temperature. This will lead to increased risks to human health, a rise in the sea level, and other adverse changes to plant and animal habitat.

Toxic Chemicals—In the air, NOx reacts readily with common organic chemicals and even ozone, to form a wide variety of toxic products, some of which may cause biological mutations. Examples of these chemicals include the nitrate radical, nitroarenes, and nitrosamines.

Visibility Impairment—Nitrate particles and nitrogen dioxide can block the transmission of light, reducing visibility in urban areas and on a regional scale in our national parks.

PM: WHAT IS IT? WHERE DOES IT COME FROM?

Particulate matter, or PM, is the term for particles found in the air, including dust, dirt, soot, smoke, and liquid droplets. Particles can be suspended in the air for long periods of time. Some particles are large or dark enough to be seen as soot or smoke. Others are so small that individually they can only be detected with an electron microscope.

Some particles are directly emitted into the air. They come from a variety of sources such as cars, trucks, buses, factories, construction sites, tilled fields, unpaved roads, stone crushing, and burning of wood.

Other particles may be formed in the air from the chemical change of gases. They are indirectly formed when gases from burning fuels react with sunlight and water vapor. These can result from fuel combustion in motor vehicles, at power plants, and in other industrial processes.

Chief Causes for Concern
PM

- is associated with serious health effects.

- is associated with increased hospital admissions and emergency room visits for people with heart and lung disease.

- is associated with work and school absences.

- is the major source of haze that reduces visibility in many parts of the United States, including our National Parks.

- settles on soil and water and harms the environment by changing the nutrient and chemical balance.

- causes erosion and staining of structures including culturally important objects such as monuments and statues.

Health problems for sensitive people can get worse if they are exposed to high levels of PM for several days in a row.

Health and Environmental Impacts of PM
Particulate matter causes a wide variety of health and environmental impacts.

Health Effects
- Many scientific studies have linked breathing PM to a series of significant health problems, including:

  — aggravated asthma

— increases in respiratory symptoms like coughing and difficult or painful breathing

— chronic bronchitis

— decreased lung function

— premature death.

Visibility Impairment

- PM is the major cause of reduced visibility (haze) in parts of the United States, including many of our national parks.

Atmospheric Deposition

- Particles can be carried over long distances by wind and then settle on ground or water. The effects of this settling include:

  — making lakes and streams acidic

  — changing the nutrient balance in coastal waters and large river basins

  — depleting the nutrients in soil

  — damaging sensitive forests and farm crops

  — affecting the diversity of ecosystems.

Aesthetic Damage

- Soot, a type of PM, stains and damages stone and other materials, including culturally important objects such as monuments and statues.

SO_2: WHAT IS IT? WHERE DOES IT COME FROM?

Sulfur dioxide, or SO_2, belongs to the family of sulfur oxide gases (SOx). These gases dissolve easily in water. Sulfur is prevalent in all raw materials, including crude oil, coal, and ore that contains common metals like aluminum, copper, zinc, lead, and iron. SOx gases are formed when fuel containing sulfur, such as coal and oil, is burned, and when gasoline is extracted from oil, or metals are extracted from ore. SO_2 dissolves in water vapor to form acid, and interacts with other gases and particles in the air to form sulfates and other products that can be harmful to people and their environment.

Over 65% of SO_2 released to the air, or more than 13 million tons per year, comes from electric utilities, especially those that burn coal. Other sources of SO_2 are industrial facilities that derive their products from raw materials like metallic ore, coal, and crude oil, or that burn coal or oil to produce process heat. Examples are petroleum refineries, cement manufacturing, and metal processing facilities. Also, locomotives, large ships, and some nonroad diesel equipment currently burn high sulfur fuel and release SO_2 emissions to the air in large quantities.

Chief Causes for Concern

SO_2 contributes to respiratory illness, particularly in children and the elderly, and aggravates existing heart and lung diseases.

SO_2 contributes to the formation of acid rain, which:

- damages trees, crops, historic buildings, and monuments; and

- makes soils, lakes, and streams acidic.

SO_2 contributes to the formation of atmospheric particles that cause visibility impairment, most noticeably in national parks.

SO_2 can be transported over long distances.

SO_2 and the pollutants formed from SO_2, such as sulfate particles, can be transported over long distances and deposited far from the point of origin. This means that problems with SO_2 are not confined to areas where it is emitted.

Short-term Peak Levels

High levels of SO_2 emitted over a short period, such as a day, can be particularly problematic for people with asthma. EPA encourages communities to learn about the types of industries in their communities and to work with local industrial facilities to address pollution control equipment failures or process upsets that could result in peak levels.

Health and Environmental Impacts of SO_2

SO_2 causes a wide variety of health and environmental impacts because of the way it reacts with other substances in the air. Particularly sensitive groups include people with asthma who are active outdoors and children, the elderly, and people with heart or lung disease.

Respiratory Effects From Gaseous SO_2

Peak levels of SO_2 in the air can cause temporary breathing difficulty for people with asthma who are active outdoors. Longer-term exposures to high levels of SO_2 gas and particles cause respiratory illness and aggravate existing heart disease.

Respiratory Effects From Sulfate Particles

SO_2 reacts with other chemicals in the air to form tiny sulfate particles. When these are breathed, they gather in the lungs and are associated with increased respiratory symptoms and disease, difficulty in breathing, and premature death.

Visibility Impairment

Haze occurs when light is scattered or absorbed by particles and gases in the air. Sulfate particles are the major cause of reduced visibility in many parts of the U.S., including our national parks.

Acid Rain

SO_2 and nitrogen oxides react with other substances in the air to form acids, which fall to earth as rain, fog, snow, or dry particles. Some may be carried by the wind for hundreds of miles.

Plant and Water Damage

Acid rain damages forests and crops, changes the makeup of soil, and makes lakes and streams acidic and unsuitable for fish. Continued exposure over a long time changes the natural variety of plants and animals in an ecosystem.

Aesthetic Damage

SO_2 accelerates the decay of building materials and paints, including irreplaceable monuments, statues, and sculptures that are part of our nation's cultural heritage.

CO: WHAT IS IT? WHERE DOES IT COME FROM?

Carbon monoxide, or CO, is a colorless, odorless gas that is formed when carbon in fuel is not burned completely. It is a component of motor vehicle exhaust, which contributes about 56 percent of all CO emissions nationwide. Other non-road engines and vehicles (such as construction equipment and boats) contribute about 22 percent of all CO emissions nationwide. Higher levels of CO generally occur in areas with heavy traffic congestion. In cities, 85 to 95 percent of all CO emissions may come from motor vehicle exhaust. Other sources of CO emissions include industrial processes (such as metals processing and chemical manufacturing), residential wood burning, and natural sources such as forest fires. Woodstoves, gas stoves, cigarette smoke, and unvented gas and kerosene space heaters are sources of CO indoors. The highest levels of CO in the outside air typically occur during the colder months of the year when inversion conditions are more frequent. The air pollution becomes trapped near the ground beneath a layer of warm air.

Chief Causes for Concern

CO

- is poisonous even to healthy people at high levels in the air.

- can affect people with heart disease.

- can affect the central nervous system.

Motor Vehicle Use Is Increasing

Nationwide, three-quarters of carbon monoxide emissions

come from on-road motor vehicles (cars and trucks) and non-road engines (such as boats and construction equipment). Control measures have reduced pollutant emissions per vehicle over the past 20 years, but the number of cars and trucks on the road and the miles they are driven have doubled in the past 20 years. Vehicles are now driven two trillion miles each year in the United States. With more and more cars traveling more and more miles, growth in vehicle travel may eventually offset progress in vehicle emissions controls.

Malfunctions and Tampering Reduce the Effectiveness of Emission Control Systems

Today's sophisticated emission control systems on vehicles are designed to keep pollution to a minimum, but vehicles quickly become polluters when their emission controls do not work correctly or if drivers tamper with them.

Health and Environmental Impacts of CO

Carbon monoxide can cause harmful health effects by reducing oxygen delivery to the body's organs (like the heart and brain) and tissues.

Cardiovascular Effects The health threat from lower levels of CO is most serious for those who suffer from heart disease, like angina, clogged arteries, or congestive heart failure. For a person with heart disease, a single exposure to CO at low levels may cause chest pain and reduce that person's ability to exercise; repeated exposures may contribute to other cardiovascular effects.

Central Nervous System Effects Even healthy people can be affected by high levels of CO. People who breathe high levels of CO can develop vision problems, reduced ability to work or learn, reduced manual dexterity, and difficulty performing complex tasks. At extremely high levels, CO is poisonous and can cause death.

Smog
CO contributes to the formation of smog ground-level ozone, which can trigger serious respiratory problems.

LEAD: WHAT IS IT? WHERE DOES IT COME FROM?
Lead is a metal found naturally in the environment as well as in manufactured products. The major sources of lead emissions have historically been motor vehicles (such as cars and trucks) and industrial sources. Due to the phase out of leaded gasoline, metals processing is the major source of lead emissions to the air today. The highest levels of lead in air are generally found near lead smelters. Other stationary sources are waste incinerators, utilities, and lead-acid battery manufacturers.

Chief Causes for Concern
Lead
- particularly affects young children and infants

- is still found at high levels in urban and industrial areas

- deposits on soil and water and harms animals and fish.

Children Are at Greatest Risk
Although overall blood lead levels have decreased since 1976, infants and young children still have the highest blood lead levels. Children and others can be exposed to lead not only through the air, but also through accidentally or intentionally eating soil or paint chips, as well as food or water contaminated with lead.

High Levels of Lead Are Still of Concern in Localized Areas
Urban areas with high levels of traffic, trash incinerators, or other industry, as well as areas near lead smelters, battery plants, or industrial facilities that burn fuel, may still have high lead levels in air. In 1999, ten areas of the country did not meet the national health-based air quality standards for lead.

Health and Environmental Impacts of Lead

People, animals, and fish are mainly exposed to lead by breathing and ingesting it in food, water, soil, or dust. Lead accumulates in the blood, bones, muscles, and fat. Infants and young children are especially sensitive to even low levels of lead.

Damages Organs   Lead causes damage to the kidneys, liver, brain and nerves, and other organs. Exposure to lead may also lead to osteoporosis (brittle bone disease) and reproductive disorders.

Affects the Brain and Nerves   Excessive exposure to lead causes seizures, mental retardation, behavioral disorders, memory problems, and mood changes. Low levels of lead damage the brain and nerves in fetuses and young children, resulting in learning deficits and lowered IQ.

Affects the Heart and Blood   Lead exposure causes high blood pressure and increases heart disease, especially in men. Lead exposure may also lead to anemia, or *weak blood*.

Affects Animals and Plants   Wild and domestic animals can ingest lead while grazing. They experience the same kind of effects as people who are exposed to lead. Low concentrations of lead can slow down vegetation growth near industrial facilities.

Affects Fish   Lead can enter water systems through runoff and from sewage and industrial waste streams. Elevated levels of lead in the water can cause reproductive damage in some aquatic life and cause blood and neurological changes in fish and other animals that live there.

WHAT ARE TOXIC AIR POLLUTANTS?

Toxic air pollutants, also known as hazardous air pollutants, are those pollutants that are known or suspected to cause cancer or other serious health effects, such as reproductive effects or birth defects, or adverse environmental effects. EPA is working with state, local, and tribal governments to reduce air toxics releases of 188 pollutants to the environment. Examples of toxic air pollutants include benzene, which is found in gasoline; perchlorethlyene, which is emitted from some dry cleaning facilities; and methylene chloride, which is used as a solvent and paint stripper by a number of industries. Examples of other listed air toxics include dioxin, asbestos, toluene, and metals such as cadmium, mercury, chromium, and lead compounds.

What Are the Health and Environmental Effects of Toxic Air Pollutants?

People exposed to toxic air pollutants at sufficient concentrations and durations may have an increased chance of getting cancer or experiencing other serious health effects. These health effects can include damage to the immune system, as well as neurological, reproductive (e.g., reduced fertility), developmental, respiratory and other health problems. In addition to exposure from breathing air toxics, some toxic air pollutants such as mercury can deposit onto soils or surface waters, where they are taken up by plants and ingested by animals and are eventually magnified up through the food chain. Like humans, animals may experience health problems if exposed to sufficient quantities of air toxics over time.

Where Do Toxic Air Pollutants Come From?

Most air toxics originate from human-made sources, including mobile sources (e.g., cars, trucks, buses) and stationary sources (e.g., factories, refineries, power plants), as well as indoor sources (e.g., some building materials and cleaning solvents). Some air toxics are also released from natural sources such as volcanic eruptions and forest fires.

How Are People Exposed to Air Toxics?

People are exposed to toxic air pollutants in many ways that can pose health risks, such as by:

- Breathing contaminated air.

- Eating contaminated food products, such as fish from contaminated waters; meat, milk, or eggs from animals that fed on contaminated plants; and fruits and vegetables grown in contaminated soil on which air toxics have been deposited.

- Drinking water contaminated by toxic air pollutants.

- Ingesting contaminated soil. Young children are especially vulnerable because they often ingest soil from their hands or from objects they place in their mouths.

- Touching (making skin contact with) contaminated soil, dust, or water (for example, during recreational use of contaminated water bodies).

Once toxic air pollutants enter the body, some persistent toxic air pollutants accumulate in body tissues. Predators typically accumulate even greater pollutant concentrations than their contaminated prey. As a result, people and other animals at the top of the food chain who eat contaminated fish or meat are exposed to concentrations that are much higher than the concentrations in the water, air, or soil.

Can I Find out About the Toxics in My Community?

- **National Air Toxics Assessment**—This site provides emissions and health risk information on 33 air toxics that present the greatest threat to public health in the largest number of urban areas. Maps and lists are available and can be requested by state or county level.

- **Toxics Release Inventory**—This database includes information for the public about releases of toxic chemicals from manufacturing facilities into the environment through the air, water, and land. You can access the data by typing in your zip code.

Is Air Pollution a Factor in Heart Disease?

In a July 2004 press release, the U.S. Environmental Protection Agency (EPA) announced a $30 million grant award to the University of Washington to study the connection between air pollution and cardiovascular (heart) disease. EPA administrator Mike Leavitt explained that an important component in the national strategy to improve U.S. air quality is "to improve our understanding of the health risks from long-term exposure to particulate pollution, particularly as it relates to heart disease, the leading cause of death in our country."[1]

It was not until May 2004 that the American Heart Association (AHA) issued an expert statement about the effects of exposure to air pollutants. Based on scientific reports from the previous decade, the AHA stated that air pollution is a serious public health problem. The following article identifies the connection between air pollution and heart disease.

Air pollution contains different components. The EPA has identified and set air quality standards for six common air contaminants: ground-level ozone, nitrogen oxides, particulate matter, sulfur dioxide, carbon monoxide, and lead. There is additional concern about the effect of indoor air pollution on people's health. One indoor pollutant is secondhand tobacco smoke, which is potentially damaging to coronary (heart) functions.

The EPA has developed an Air Quality Index that provides daily information for more than 150 cities around the country. This daily report can be found online at *www.epa.gov/airnow*. The EPA has been working on the Clean Air Rules of 2004, a series of regulations that address different aspects of air pollution. One problem is enforcement: As the following report states, 19% of all U.S. counties with air quality monitoring systems are not meeting EPA air quality standards. In some states or regions of the country, the percentage is much higher. With the EPA considering imposing stricter standards and enforcement, the cooperation of more industries, and increased state involvement are moving this country toward cleaner air.

—The Editor

1. U.S. Environmental Protection Agency. Press Release. *EPA Awards Largest-Ever Grant to Study Health Effects of Air Pollution*. July 2004. Available online at *http://www.epa.gov/epahome/newsroom.htm*.

The Effects of Air Pollution on Health
from the American Heart Association

Exposure to air pollution contributes to the development of cardiovascular diseases (heart disease and stroke).

A person's relative risk due to air pollution is small compared with the impact of established cardiovascular risk factors such as smoking, obesity, or high blood pressure. However, this is a serious public health problem because an enormous number of people are exposed over an entire lifetime.

BACKGROUND

Until May of 2004, the American Heart Association had not issued any expert reviewed statement about the short-term and long-term effects of chronic exposure to different pollutants. This was due to flaws in research design and methodology of many pollution studies. During the last decade, however, epidemiological studies conducted worldwide have shown a consistent, increased risk for cardiovascular events, including heart and stroke deaths, in relation to short- and long-term exposure to present-day concentrations of pollution, especially particulate matter.

Elderly patients, people with underlying heart or lung disease, lower socioeconomic populations and diabetics may be at particularly increased risk. More research is needed to find out the differential toxicity of various constituents of air pollution.

COMPONENTS OF AIR POLLUTION

Air pollution is composed of many environmental factors. They include carbon monoxide, nitrates, sulfur dioxide, ozone, lead, secondhand tobacco smoke and particulate matter. Particulate matter, also known as particle pollution, is composed of solid and liquid particles within the air. It can be generated from vehicle emissions, tire fragmentation and road dust, power generation and industrial combustion, smelting and other metal processing, construction and demolition activities, residential wood burning, windblown soil, pollens, molds, forest fires, volcanic emissions and sea spray. These particles vary considerably in size, composition and origin.

PARTICULATE MATTER AND SULFUR DIOXIDE

The concentrations of both particulate matter and sulfur dioxide often change in parallel. The oxidation of sulfur dioxide in the atmosphere is linked with the formation of various particulate compounds, including acid sulfates.

A 1994 report on the adverse effects of particulate air pollution, published in the *Annual Reviews of Public Health*, noted a 1 percent increase in total mortality for each 10 mg/m³ increase in particulate matter. Respiratory mortality increased 3.4 percent and cardiovascular mortality increased 1.4 percent. More recent research suggests that one possible link between acute exposure to particulate matter and sudden death may be related to sudden increases in heart rate or changes in heart rate variability.

The Environmental Protection Agency (EPA) has declared that "tens of thousands of people die each year from breathing tiny particles in the environment." A recent report released by the nonprofit Health Effects Institute in Cambridge, Mass., agrees with the EPA assessment. This study was reviewed by *Science* magazine and clearly shows that death rates in the 90 largest U.S. cities rise by 0.5 percent with only a tiny increase— 10 micrograms (mcg) per cubic meter—in particles less than 10 micrometers in diameter. This finding is similar to those of other studies throughout the world. The case is stronger with this study, because it eliminated several factors that could confound the interpretation of the data, such as temperature and other pollutants.

The number of deaths due to cardiac and respiratory problems may be small when looking at individual cities with small particles in the environment. The combined long-term effect of studies in several large cities predicts 60,000 deaths each year caused by particulate matter. This is a staggering loss of life that can be eliminated by stricter emissions standards as proposed by the EPA.

SECONDHAND TOBACCO SMOKE

Secondhand smoke, also known as environmental tobacco smoke, is the single largest contributor to indoor air pollution when a smoker is present. Studies of secondhand smoke indicate that air pollution in general can affect the heart and circulatory system. Previous research has established that exposure to the secondhand smoke of just one cigarette per day accelerates the progression of atherosclerosis—thus it is plausible that even low doses of air pollution could negatively affect coronary functions.

CARBON MONOXIDE

Carbon monoxide (CO) is a colorless, odorless and highly poisonous gas. It's a common air pollutant associated with combustion reactions in cars and other vehicles. It's also in cigarette smoke. When the level of CO in blood increases, the level of oxygen that blood can carry decreases. That's why CO in any level is harmful to your body—and high levels may prove deadly. Long-term, low-level exposure to carbon monoxide may lead to serious respiratory diseases. Smoking tobacco and breathing environmental tobacco smoke raise CO levels in your blood, eventually leading to disease.

Carbon monoxide levels in the blood of nonsmokers vary depending on the quality of air that they generally breathe. The levels are usually 0–8 parts per million (abbreviated ppm). The CO level of smokers is much higher, but it depends on when and how much they smoke, and how they smoke (cigar, pipe, cigarette, etc.). A person who smokes one pack of cigarettes a day has a blood CO level of 20 ppm; someone who smokes two packs a day may have a blood CO level of 40 ppm. When smoking stops, the blood CO level should return to normal in a few days. . . .

Nicotine causes a short-term increase in blood pressure, heart rate and blood flow from the heart. It can also cause arteries to narrow. The carbon monoxide reduces the amount of oxygen the blood can carry. When combined with nicotine's effects, this creates an imbalance between cells' increased demand for oxygen and the reduced amount of oxygen that the blood can supply.

NITROGEN DIOXIDE

Nitrogen dioxide (NO_2) is a precursor to ozone (O_3) formation. Current efforts to reduce ozone levels also target reductions in NO_2 levels. In contrast to ozone, NO_2 is often found at higher levels indoors compared with outdoors. Mainly this occurs in settings where gas stoves and kerosene heaters are being used.

The main sources of NO and NO_2 in outdoor air are emissions from vehicles and from power plants and other fossil fuel-burning industries. NO_2 levels vary with traffic density. Annual average concentrations range from 0.015–0.035 ppm. Some highly congested areas like metropolitan Los Angeles ranged from 0.020–0.056 ppm in 1990. Estimates of concentrations inside vehicles in Los Angeles ranged from 0.028–0.078 ppm, where average commuting time was about 6.5 hours per week.

People with respiratory or heart problems should avoid prolonged exposure to high-traffic areas and unventilated heating elements in their homes.

People with asthma appear to be especially vulnerable to the effects of acute NO_2 exposure. Healthy people, by contrast, don't seem to show detectable changes in lung function. Exposure to high levels (20 ppm) for several weeks or longer causes emphysema-like changes in the lungs of animals.

EPA AIR QUALITY STANDARDS

The U.S. Environmental Protection Agency (EPA) introduced its 1997 National Ambient Air Quality Standards (NAAQS) to educate the public about daily air quality levels, including information about ozone and particulate matter levels. This daily Air Quality Index was updated in 2003 to include information on fine particle pollution. This index provides information each day for more than 150 cities along with a health alert system that reflects recommended changes in activity on days when pollution is high. These daily updates can be found on the EPA Website at *www.epa.gov/airnow* and in many newspapers across the country.

The American Heart Association supports these EPA guidelines for activity restriction for people with heart disease or those who have certain cardiovascular risk factors and for people with pulmonary disease and diabetes and the elderly.

Nineteen percent of all U.S. counties with air-quality monitoring systems are presently not meeting these standards. This inadequacy soars to much higher estimates in regions such as the industrial Midwest (41 percent) and California (60 percent).

AMERICAN CANCER SOCIETY COHORT STUDY

Recently published data from the American Cancer Society cohort suggested that long-term exposure to fine particulate air pollution at levels that occur in North America is associated with increased risk for cardiovascular mortality by 12 percent for every 10 micrometers of particulate matter within 1 cubic meter of air. Ischemic heart diseases (e.g., heart attacks) account for the largest portion of this increased mortality rate. Other causes, such as heart failure and fatal arrhythmias, also increased.

AIR POLLUTION IMPACT IN U.S. CITIES

Another study confirmed the importance of variations of pollution within a single city. Its findings suggested that a person's exposure to toxic components of air pollution may vary as much within one city as across different cities. After studying 5,000 adults for eight years, the researchers also found that exposure to traffic-related air pollutants was more highly related to mortality than were city-wide background levels. For example, those who lived near a major road were more likely to die of a cardiovascular event.

Some research has estimated that people living in the most polluted U.S. cities could lose between 1.8 and 3.1 years because of exposure to chronic air pollution. This has led some scientists to conclude that

1. Short-term exposure to elevated levels of particle pollution is associated with a higher risk of death due to a cardiovascular event.

2. Hospital admissions for several cardiovascular and pulmonary diseases rise in response to higher concentrations of particle pollution.

3. Prolonged exposure to elevated levels of particle pollution is a factor in reducing overall life expectancy by a few years.

Does Air Pollution Contribute to Asthma and Respiratory Problems?

How much air do you breathe each day? An adult takes in about 3,000 gallons of air every day. Children breathe even more. When there are pollutants in that air, health problems result. The following article outlines some scientific research that links cases of asthma to air pollution. The problem is of such significance that in May 2004 the Environmental Protection Agency (EPA) launched Asthma Awareness Month, to help the 20 million Americans who suffer from asthma. The number of sufferers includes 6.3 million children.[1]

So what makes up this dirty air we call "smog"? Smog is mostly ozone. Ozone found too close to the Earth's surface causes health problems, but when it blankets the Earth about 20 miles [32 km] above, it is beneficial, protecting us from radiation. Smog forms in the air when volatile organic compounds (VOCs) and nitrogen oxides (NOx) react to the sun's heat and light. A major source of these smog-forming chemicals are cars and industrial smokestacks. Another component of smog is particulate matter, or particles from burning wood or fuels, dust, dirt, and industrial plants. Particulate matter adds to the brown haze of smog. It can blow miles from where it originally forms, which allows the problem to easily cross city and state boundaries.

—The Editor

1. U.S. Environmental Protection Agency. EPA Newsroom. "May Is Allergy Awareness Month." May 2004. Available online at *http://www.epa.gov/ newsroom/allergy_month.htm.*

Is Air Pollution Making Us Sick?
by Kimi Eisele

Every Day, Thousands of Asthmatics Are Taken to the Emergency Room. As Asthma Becomes a National Epidemic, People Want to Know: Is Air Pollution Making Us Sick?

When I was eight years old, I bit into a red apple and my throat closed. It was an autumn morning, and I was sitting in the back seat of the car. My mouth began to itch and I felt like I had to burp, but I couldn't. I could barely speak. Outside, dry leaves were scuttling across the sidewalk. I couldn't get enough air. My parents, twisting around in alarm to look at me from the front seat, started inhaling and exhaling deeply and slowly, as if modeling breathing would remind me how to do it. But I hadn't forgotten how to breathe. I simply couldn't.

Soon afterward, I was diagnosed with atopic asthma—asthma caused by allergies, which is a chronic condition with no known cure. The word *asthma* comes from the Greek word for "panting," which is what we asthmatics do when we're trying to get air. My own panting is induced by any number of factors: dust mite dung, dry wind, cat dander, cold air, rabbits, wood smoke, pollen, guinea pigs, cigarettes, grass, horses, and some species of trees, among other things. Any of these will make my eyes itch, my nose run, and my skin break out in a rash. Or they will shut my airways.

Because I have asthma, for most of my life I've had to pay attention to my surroundings. I've had to be aware of hovering dust, the direction of a spring breeze, the presence of a cat. Luckily, my asthma is mild. My attacks aren't life-threatening, and as long as I have my inhaler along, they're easily relieved. And I can usually get myself out of the path of the allergens, away from the cats and rabbits, in from the pollens, and into a "clean" air space.

But for the 17 million other people with asthma in the United States, controlling the air space is not always so simple, especially because some of the most common triggers of asthma—smog and soot from tailpipe exhaust and power plants—are in the air we breathe every day.

It's possible that air pollution is doing more than just triggering asthma attacks. It may also be an element in the development of the disease—a criminal accomplice, not just an accessory after the fact. In industrialized countries, asthma is becoming more common and more severe. Five thousand people die of it every year in the United States. Currently it's the sixth most common chronic condition in the nation. Three times as many people have it now as in 1980. Some 6 million of them are children. For children, asthma is the most common chronic disorder, the leading cause of missed school, and the leading cause of hospitalization.

Is polluted air helping to drive this epidemic? As yet, there's no scientific consensus. But the evidence pointing to air pollution as one major culprit is getting harder and harder to ignore.

A seventeenth-century Belgian physician and asthmatic named Jean Baptista van Helmont called asthma the "falling sickness of the lungs." I know what he meant. As an allergic asthmatic, I am hypersensitive to substances that may have no effect at all on someone standing next to me. All it takes is a speck, a drop, a few molecules, and my immune system responds as if I were being attacked by something virulent. T cells, macrophages, eosinophils, and other immune system cells flood into the tissues of my airways, setting off an immunological train wreck. As my airways become inflamed and swollen and start producing mucus, I cough, drip, and wheeze. The immune system cells also prompt the most frightening response—constriction of the smooth muscle of my bronchial tubes. This is van Helmont's "falling" effect. It feels as if my airways were collapsing.

What makes the asthmatic's airways so hypersensitive? Genetics plays a part; I probably inherited my asthma from my asthmatic father. But DNA isn't the whole story. "Genetic changes haven't occurred rapidly enough to account for the global increase in asthma," says Anne Wright of the Arizona Respiratory Center at the University of Arizona in Tucson. "It pretty clearly has something to do with our interaction with the environment."

One theory suggests that modern society may be too antiseptic for our bodies. Over the last hundred years, humans' relationship to microorganisms has changed dramatically. When people were born in homes, didn't bathe as regularly, and didn't use antibiotics, they encountered more germs. Children got more sick more often. But exposure to microbes may have forced children's immune systems into maturity—and strengthened them against exposure to allergens. In one study following 1,200 people for more than two decades, Wright and her colleagues found that those who attended day care as children, or whose family owned a dog when they were growing up, tended to have less asthma than others. What do dogs and day care have in common? Bacteria—mostly from feces, Wright explains. "We all know poop's not good for you," she says. "You can get sick from it. But what also happens is that your immune system is developing antibodies and being activated by those organisms."

Although Wright subscribes to the hygiene hypothesis, she and other researchers agree that asthma, like cancer, has no single cause. "We won't find treatments and cures and preventive measures if we don't address the disease from different points of view—environmental, genetic, molecular, biological," says Fernando Martinez, director of the Arizona Respiratory Center. Children and adults who have been frequently exposed to tobacco smoke and to indoor allergens from cats, cockroaches, and household dust mites have more asthma than those who haven't. The disease is more prevalent and more severe among poor people. It's more common in inner cities. Stress may cause asthma. Sedentary lifestyles and unhealthy diets have also been associated with asthma.

No wonder asthma has seemed exhaustingly complicated all my life. Even the suggested remedies are overwhelming—everything from smoking marijuana to giving up dairy products to inhaling steroids to refraining from sex. And the mystery of asthma's causes has always left me asking, "Why me?" Why does standing near a horse throw my lungs into a wheezing fit, and not the next guy's?

On pollution, scientists are making progress. Several long-term, multimillion-dollar studies are now underway to track children from the womb onward, measuring precisely what contaminants they're exposed to and recording who develops asthma and who doesn't. One study will investigate effects of the mix of air pollutants and pesticides that descends on children in central California. In the next few years, this research should start to shape concrete answers.

And last February [2002], researchers from the University of Southern California published the most persuasive evidence yet linking asthma and air pollution. The study followed more than 3,500 children from twelve Southern California communities, six of which endured the kind of smog for which the Los Angeles region is notorious, and six of which had fairly clean air. Smog's primary ingredient is ozone, a caustic gas formed when sunlight and heat acts on certain air pollutants—namely, nitrogen oxides and hydrocarbons. In Southern California, by far the largest source of these pollutants is tailpipes.

None of the children had asthma when the study began. After five years, 265 were diagnosed with it. But the critical finding was that children who lived in high-ozone areas and were involved in several team sports were three times more likely to develop asthma than couch potatoes living in less polluted communities. "Kids playing three or more sports are likely to be outdoors ventilating at high rates, and are therefore being exposed to higher levels of air pollution," explains James Gauderman, one of the study's authors.

But it's only a beginning. Martinez is one researcher who says the findings are important but not conclusive. High rates of asthma in cities may be related to factors such as stress, he argues. And he points out that asthma comes in different varieties. In the days when East Germany was highly polluted, its population had higher rates of asthma—but fewer allergies— than West Germans. It's possible, Martinez says, that pollution is not a risk factor for the allergic form of the disease, but may be a factor for another form of the disease.

Phoenix, Arizona, a fast-growing, sprawling desert city, is one of the most ozone-polluted cities in the country. In mid-September, not long after dawn, I drive there to visit the Breathmobile, a mobile asthma clinic of the Phoenix Children's Hospital. Though the desert air feels crisp at this hour, from the highway I can see the thick green-gray stripe of smog that has already spread across the horizon. On especially polluted days, flashing signs on the road are turned on to warn drivers that the air is bad.

Children in Phoenix's urban schools have a lot of asthma. Maricopa County, where the city is located, has one of the highest death rates of asthma in the nation—2.1 percent in 1999.

This morning the Breathmobile is at William T. Machan Elementary School. To get there, I drive through a residential neighborhood with tree-lined streets, single-story homes, and semi-green lawns. Though it doesn't fit the stereotype, this is an inner-city school. Downtown Phoenix and its heavy automobile traffic are only blocks away. Students here face the same challenges that children in most inner cities in America face: low family incomes, poor access to health care, and abnormally high rates of asthma.

The Breathmobile is a large motor home. On the outside, drawings of smiling children and large colorful letters advertise the Phoenix Children's Hospital and the project's main sponsor, Wal-Mart. Inside, it looks very much like a doctor's office. Children referred by teachers and school nurses are tested here for pulmonary function on a computer that measures the force of air exhaled from their lungs. Just asking children whether they're having trouble breathing isn't enough, explains Judy Harris, director of the Breathmobile. "Often the actual function of their lungs is different from what kids say they feel, because they're used to living with low-level asthma," she says. If asthma is diagnosed, the children and their parents get free medicine and training in asthma management.

Fifth-grader Elizabeth Vargas, back for a checkup, is the first patient of the day. Her lungs don't look good. Each time she exhales into a tube connected to the computer, its screen shows a picture of a balloon being blown up. A healthy child would be able to pop the balloon. But until Elizabeth inhales the bronchiodilator medication albuterol, she can pop only two out of five balloons. After her appointment, Elizabeth will take home a supply of Advair, a steroid-based preventive daily medication. Unless her asthma becomes less severe as she grows up, as sometimes happens, she may have to take it for the rest of her life.

Elizabeth is a typical Breathmobile patient. Like her, 75 percent of the Breathmobile's patients are Hispanic. Also like her, many of them get their first real medical help here. Elizabeth has suffered from tight lungs all her life, so much so that when Harris asks in Spanish what happens when she runs, she answers, "My lungs get agitated, and it hurts here"—putting a hand to the center of her chest. She missed thirty days of school last winter because of asthma. Yet her condition went undiagnosed until her first visit to the clinic several months ago.

When asthma gripped my own lungs, I had the middle-class advantages of good air—we lived on a quiet street in a small college town—and good health care. My parents even bought a special vacuum cleaner that sucked dust into a vat of water so it wouldn't blow back into the house. But today, those who lack such socio-economic cushions suffer disproportionately from asthma. This fact infuriates many environmental justice advocates, who believe the link between asthma and air pollution is as obvious as the sooty air in inner-city neighborhoods.

"It's certainly no accident that the neighborhoods with the highest rates of asthma also have a high incidence of polluting facilities, and that they're also low-income communities of color," says Omar Freilla. Freilla works at Sustainable South Bronx, a New York City group that seeks to reduce pollution and promote parks in one of the city's most environmentally blighted areas. The South Bronx, he points out, is the site of twenty-six waste facilities and the largest food distribution center in the world. The number of trucks passing through the food distribution center's neighborhood has been estimated at 11,000 daily. And Freilla believes that the asthma hospitalization rate in the South Bronx—six times higher than the national average—is directly related to all those tailpipes.

Freilla might get a sympathetic hearing if he visited the Tokyo District Court. This October, the court handed seven Tokyo asthmatics a victory in their lawsuit against the national government and the highway authority. Tailpipe pollution in the city, the judges said, had "caused and exacerbated" the plaintiffs' disease.

In Tokyo, as in the Bronx, the worst problem is diesel. A contributor to ozone and a source of unhealthy substances such as sulfur dioxide, diesel exhaust also carries soot in particles so tiny as to be invisible. Breathing fine soot is extremely dangerous. When diesel particulate levels in the air go up, so do death rates among the sick and elderly. And so do the number of asthmatics admitted to emergency rooms.

What about development of the disease? Andrew Saxon, chief of clinical immunology at the University of California at Los Angeles, exposed volunteers to an allergen they were never likely to encounter in the normal course of events, a mollusk protein called KLH. Initially, none of the volunteers showed any sensitivity at all to the protein. But when KLH was combined with diesel exhaust particles, people who had breathed it easily beforehand demonstrated a heightened allergic response. In another study, Saxon combined known allergens with diesel exhaust and exposed allergic volunteers to the combination. He saw a fivefold increase in total allergic protein levels and a fifty-fold increase in the allergic antibodies. In his experiments, the dosage of diesel exhaust particles was 0.3 milligrams—comparable to what you'd breathe in two days in Los Angeles or one day in Tokyo.

"Something about diesel exhaust primes the immune system," explains Gina Solomon, a doctor and medical researcher at NRDC. "Even a mild dose of common allergen that you might not have noticed suddenly turns into a major reaction."

Along with researchers at the University of California at San Francisco, Solomon analyzed dozens of recent studies on diesel for an article in the peer-reviewed journal *Environmental Health Perspectives.* They found epidemiological studies showing, among other things, that children living near heavily traveled trucking routes are much more likely than others to develop wheezing and have trouble breathing.

But most interesting were the studies, that, like Saxon's, point to a physiological connection between diesel and allergies. Diesel soot steps up the body's production of many of the immune system cells and antibodies that drive allergies and asthma attacks. Among them is the antibody called Immunoglobulin E (IgE), which is a hallmark of allergies. Diesel also raises the body's levels of TH1—a kind of immune system cell implicated in atopic asthma. Moreover, says Solomon, production of TH1 actually suppresses the body's ability to make other cells that work to prevent allergies. "So once things start to go out of whack," she says, "they can go out of whack pretty badly."

Neither Saxon nor Solomon thinks diesel is the sole answer to the asthma riddle. But unquestionably, says Saxon, "diesel is a player."

To breathe is to live. But for us asthmatics, when the air is full of pollens and particulates, what we need most becomes our worst enemy. When I was young, asthma meant only my personal affliction and the chores it demanded: regulating my breath, paying attention to allergens, and medicating. But as an adult, I'm implicated. If air pollution causes asthma, what does that mean about the energy I use and the car I drive? What does it mean that 6 million children in the United States have trouble breathing? I know I'm supposed to stay calm. But the evidence and the reality of it make me want to hyperventilate.

Michael Lerner, of the health and environmental research organization Commonweal, says the new air pollution research has "enormous political salience." He adds, "Parents are agonizingly aware of the reality that their children can't breathe, and of the tremendous impact that has on a child's life—on their ability to participate in sports and live childhood the way childhood is meant to be lived."

And then there's this: While the scientists continue their studies, the rest of us are left to control our breath. Air pollution is a hazard for many asthmatics, whether it caused our asthma or not. "We have enough data now on air pollution and asthma and mortality to say we need to be moving in the direction of more control [of pollution]," says Jonathan Samet of Johns Hopkins University. "Would another finding make a difference?"

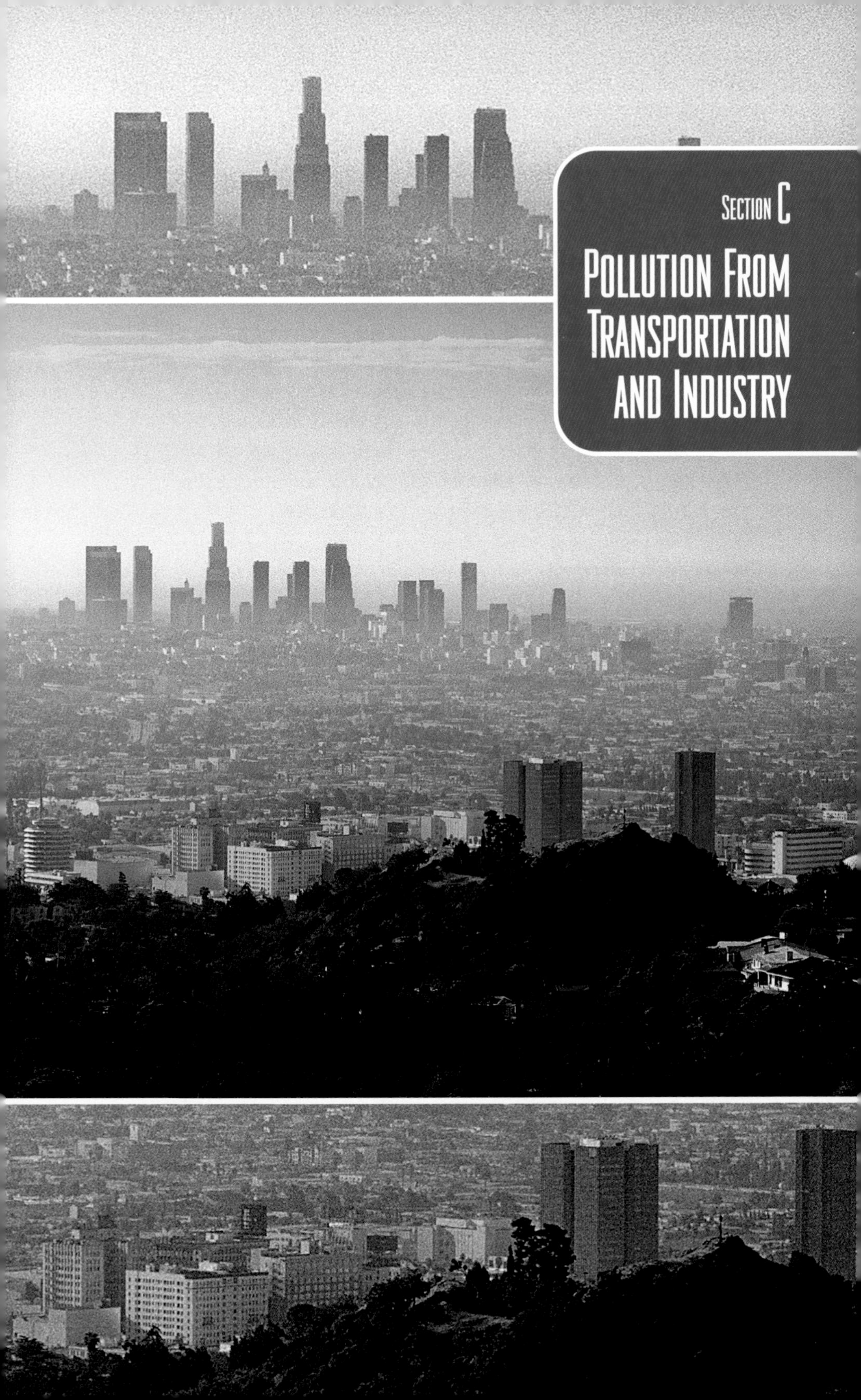

SECTION C

POLLUTION FROM TRANSPORTATION AND INDUSTRY

What Do Diesel Engines Contribute to Air Pollution?

Have you ever been stuck in traffic behind a truck spewing out black smoke? The reason the experience is not pleasant is that it is not healthy for you. In fact, scientists have found that diesel exhaust is one of the most dangerous forms of air pollution. Black diesel smoke is filled with nitrogen oxides (NOx), sulfur dioxide (SO_2), and particulate matter (PM). The particulate matter causes respiratory problems, the sulfur dioxide contributes to acid rain, and the nitrogen oxides feeds smog. The good news is that in 2001, the U.S. Environmental Protection Agency (EPA) strengthened national standards for truck and bus diesel engines to help reduce unhealthy emissions.

The following 2003 report from the American Lung Association and Environmental Defense discusses the dangers of non-road sources of diesel engines. Non-road sources include: commercial marine vessels (ships), and farm, construction, and mining equipment. All these sources contribute more sooty pollution that any other transportation source.

More good news is that in May 2004, the EPA signed the Clean Air Nonroad Diesel Rule, designed to decrease emissions from these non-road diesel sources by more that 90% and to be in full effect by 2015. As EPA administrator Mike Leavitt stated, "We're able to accomplish this in large part because of a masterful collaboration with engine and equipment manufacturers, the oil industry, state officials, and the public health and environmental communities."[1] These rules are part of a group of actions, called the Clean Air Rules of 2004, that will significantly improve the air quality in the United States.

But commercial marine engines, which are among the worst polluters, are still unregulated. According to 2004 study by the Natural Resources Defense Council, the ports of Los Angeles and Long Beach, taken together, are the largest source of air pollution in southern California.[2] In May 2004, the EPA announced a new initiative to propose new emission standards for diesel locomotives and marine diesel engines. In the meantime, some programs have already been put into place in Los Angeles (China Shipping Terminal) and New York (2003 Clean Ferry Emissions Reduction Initiative).

—The Editor

1. U.S. Environmental Protection Agency. *New Clean Diesel Rule Major Step in a Decade of Progress*. May 2004. Available online at *http://yosemite.epa.gov/opa/admpress.nsf/0/f20d2478833ea3bd85256e 91004d8f90?OpenDocument*.

2. Natural Resources Defense Council. *Harboring Pollution: Strategies to Clean Up U.S. Ports*. August 2004. Available online at *http://www.nrdc.org/air/pollution/ports/contents.asp*.

Closing the Diesel Divide
by Hilary Decker et al.

DIESEL EXHAUST IS DANGEROUS TO PUBLIC HEALTH AND THE ENVIRONMENT

The breathtaking range of hazards posed by diesel exhaust stands in stark contrast to the lack of a comprehensive approach to controlling diesel emissions from all their sources. The critical constituents of diesel exhaust include PM, NOx and SO_2, as well as a laundry list of toxic chemicals that cause both public health and environmental dangers.

What Is Diesel Exhaust?

Diesel exhaust occurs as a gas, liquid or solid and is a result of the combustion of diesel fuel in a compression-ignition engine. Its composition varies depending on the type of engine, the operating conditions, fuel characteristics and the presence of a control system, but it always contains both particulate matter and a complex mixture of hundreds of gases, many of which are known or suspected to cause cancer.

Diesel engines produce far more particulate pollution than gasoline engines. Depending on operating conditions, fuel quality and emission controls, light-duty diesel engines and heavy-duty diesel engines can emit 50 to 80 times and 100 to 200 times, respectively, more particle mass than typical catalytically equipped gasoline-powered engines.[2] Diesel particulate matter is typically fine (<2.5 microns) or ultrafine (< 0.1 micron) in size. Virtually all of the diesel exhaust particle mass has a diameter of less than 10 microns, 94 percent is less than 2.5 microns, and 92 percent is less than 1.0 microns. Because of the preponderance of small particles, diesel particulate matter is easily inhaled deep into the lungs' bronchial and alveolar regions, where their clearance is slow compared with particles deposited on airways.

More than 40 constituents of diesel exhaust are listed by either the U.S. Environmental Protection Agency or the California Air Resources Board as hazardous air pollutants or toxic air contaminants (Table 1). At least 21 of these substances are listed by the State of California as known carcinogens or reproductive toxicants.

Health Effects Specific to Diesel Exhaust

The major pollutants that make up diesel exhaust each pose threats to public health and the environment. In addition, a growing body of research on the hazards of diesel exhaust shows that this particular combination of pollutants causes significant cancer risk and both acute and chronic health problems.

Cancer risk

Numerous governmental agencies and scientific bodies have concluded that diesel exhaust is a probable human carcinogen. The first major study to investigate the contribution of diesel exhaust to people's exposures to toxic air pollutants was the Multiple Air Toxics Exposure Study (MATES-II), conducted by California's South Coast Air Quality Management District in 1998 and 1999 and one of the most comprehensive urban air toxics studies ever undertaken. The results were alarming: 70 percent of the cancer risk from air pollution for those living in the Los Angeles air basin (one of the most polluted in the country) was due to diesel particulate emissions.

Table 1: Toxic Air Contaminants and Hazardous Air Pollutants in Diesel Exhaust

Acetaldehyde*	Chlorine	Methyl ethyl ketone
Acrolein	Chlorobenzene	Naphthalene*
Aluminum	Chromium compounds*	Nickel*
Ammonia	Cobalt compounds*	4-nitrobiphenyl*
Aniline*	Copper	Phenol
Antimony compounds*	Cresol	Phosphorus
Arsenic*	Cyanide compounds	POM (including PAHs)
Barium	Dibenzofuran	Propionaldehyde
Benzene*	Dibutylphthalate	Selenium compounds*
Beryllium compounds*	Ethyl benzene	Silver
Biphenyl	Formaldehyde*	Styrene*
Bis [2-ethylhexyl]phthalate*	Hexane	Sulfuric acid
Bromine	Lead compounds*	Toluene*
1,3-butadiene*	Manganese compounds	Xylene isomers and mixtures
Cadmium*	Mercury compounds*	Zinc
Chlorinated dioxins*	Methanol	

* This compound or class of compounds is known by the State of California to cause cancer or reproductive toxicity.

See California EPA, Office of Environmental Health Hazard Assessment, "Chemicals Known to the State to Cause Cancer or Reproductive Toxicity," May 31, 2002.

Note: Toxic air contaminants on this list either have been identified in diesel exhaust or are presumed to be in the exhaust, based on observed chemical reactions or presence in the fuel or oil. See California Air Resources Board, "Toxic Air Contaminant Identification List Summaries, Diesel Exhaust," September 1997, available online at http://www.arb.ca.gov/toxics/tac/factshts/diesex.pdf.20.

As a result of this finding, the California Air Resources Board expanded the study to include all of California. The findings were similar: about 70 percent of the total inhalation

Table 2: History of Determinations of the Carcinogenicity of Diesel Exhaust

Agency	Year	Determination
National Institute for Occupational Safety and Health (NIOSH)	1988	Potential occupational carcinogen
International Agency for Research on Cancer (IARC)	1989	Probable human carcinogen
State of California	(under provision of 1990 Proposition 65)	Known by the state to cause cancer
Health Effects Institute (HEI)	1995	Potential to cause cancer
World Health Organization	1996	Probable human carcinogen
California Air Resources Board (CARB)	1998	Toxic air contaminant (determination based substantially on the cancer risk to humans)
U.S. Department of Health and Human Services National Toxicology (U.S. DHHS/NTP)	2000	Reasonably anticipated to be human carcinogen

cancer risk from air pollution for the average Californian is due to diesel exhaust, and California's Office of Environmental Health Hazard Assessment concluded that "long-term exposure

American Council of Government Industrial Hygienists (ACGIH)	2001	Suspected human carcinogen
U.S. Environmental Protection Agency (EPA)	2002	Likely human carcinogen

Sources: National Institute for Occupational Safety and Health, "Carcinogenic Effects of Exposure to Diesel Exhaust," *Current Intelligence Bulletin 50* (August 1988). Available online at *http://www.cdc.gov/niosh/88116_50.html.*

International Agency for Research on Cancer (IARC), Diesel and Gasoline Engine Exhausts and Some Nitroarenes. *IARC Monographs on the Evaluation of Carcinogenic Risks to Humans*, no. 46 (Lyons: World Health Organization, 1989), pp. 41–185.

California Environmental Protection Agency, *Chemicals Known to the State to Cause Cancer or Reproductive Toxicity* (Proposition 65, 1997), revised May 31, 2002.

Health Effects Institute, *Diesel Exhaust: A Critical Analysis of Emissions, Exposure and Health Effects* (Cambridge, MA: Health Effects Institute, 1995). Available online at *http://www.healtheffects.org/Pubs/diesum.htm, accessed on January 20, 2002.*

International Programme on Chemical Safety, World Health Organization, "Diesel Fuel and Exhaust Emissions," *Environmental Health Criteria* 171 (1996).

California's Air Resources Board, "The Toxic Air Contaminant Identification Process: Toxic Air Contaminant Emissions from Diesel-fueled Engines," fact sheet. Available online at *http://www.arb.ca.gov/toxics/dieseltac/factsht1.pdf.*

American Conference of Governmental Industrial Hygienists, "Documentation of the Threshold Limit Values and Biological Exposure Limits, Notice of Intended Changes," 2001.

U.S. Environmental Protection Agency, *Health Assessment Document for Diesel Engine Exhaust*, May 2002, EPA/600/8-90/057F.

to diesel exhaust particles poses the highest cancer risk of any toxic air contaminant evaluated. . . ." The result for the United States as a whole was even worse: 80 percent of the total cancer risk from all hazardous air pollutants is associated with the inhalation of diesel exhaust.

Acute Health Effects

Even a brief exposure to diesel exhaust can have immediate respiratory, neurological, and immunological effects. Healthy volunteers exposed to diesel exhaust for one hour showed a significant increase in airway resistance and increases in eye and nasal irritation. Other symptoms caused by exposure to diesel exhaust include coughs, headaches, light-headedness, and nausea. Epidemiological studies of bus garage workers and miners exposed to diesel exhaust on the job found decreased lung function, increased cough, labored breathing, chest tightness, and wheezing.

Chronic Non-cancer Health Effects

Long-term exposure to diesel exhaust has been associated with a greater frequency of bronchitic symptoms, cough, phlegm, and reductions in lung function. Test animals show effects including chronic inflammation of lung tissue and reduced resistance to infection, as well as significant noncarcinogenic pulmonary effects from long-term exposure.

Health Effects of Fine Particle Pollution

Because it is so laden with fine particles, diesel exhaust is implicated in all of the dangers that led EPA in 1997 to adopt stricter health-based national ambient air quality standards for fine particles. Research conducted since 1997 has confirmed EPA's findings and further documents the toll that fine particle pollution takes on our health:

- The National Morbidity, Mortality and Air Pollution Study (NMMAPS), an independent study of 90 U.S. cities using uniform methodology, reported that contemporary levels of particulate pollution are killing people. NMMAPS found strong evidence linking daily increases in particulate pollution to increases in death. In May 2002, the NMMAPS investigators at Johns Hopkins University reported an error in the software used to analyze the NMMAPS data. However, reanalysis using adjusted assumptions did not alter the main conclusions of the study (1) that there is strong evidence of an association between acute exposure to particulate air pollution and daily mortality, one day later, (2) that this association is strongest for respiratory and cardiovascular causes of death, and (3) that this association cannot be attributed to other pollutants or the weather. While some other studies of air pollution health effects have used this same software, this error does not affect the validity of the longitudinal studies on which EPA based the PM2.

Special Risks to Vulnerable Subpopulations

Children, the elderly, individuals with asthma, cardiopulmonary disease and other lung diseases, and individuals with chronic heart diseases are particularly susceptible to the effects of diesel exhaust. Air pollution affects children more than adults because they inhale more pollutants per pound of body weight and have a more rapid rate of respiration, narrower airways, and a less mature ability to metabolize, detoxify, and excrete toxins. Children also spend more time outdoors engaged in vigorous activities; athletes are similarly susceptible for this reason. Exposures that occur in childhood are of special concern because children's developmental processes can easily be disrupted and the resulting dysfunctions may be irreversible. In addition, exposures that occur early in life appear more likely to lead to disease than do exposures later in life.

NATIONAL AMBIENT AIR QUALITY STANDARDS (NAAQS)

- A study of 500,000 adults in more than 100 American cities concluded that prolonged exposure to fine particulate air pollution significantly increases the risk of dying from lung cancer and cardiopulmonary causes.

- A study of the relationship between stroke and air pollution indicates that PM10, along with the gaseous pollutants SO_2, NO_2 and CO, is a significant risk factor for acute stroke death.

- New studies and reanalysis of pre-existing work show that chronic exposure to fine particle pollution may lower life expectancy by months or years, not just by a few days.

- A 2002 Dutch study found that people living near a main road and exposed to traffic-related fine particulates and diesel soot were almost twice as likely to die from heart or lung disease and 1.4 times as likely to die of any cause compared to people living further from traffic.

- Studies consistently show a direct correlation between increased hospital admissions and increased exposure to particulate pollution.

- Evidence continues to mount that children, and particularly children with asthma, are especially sensitive to the effects of fine particle pollution.

- Increases in PM10 levels have been associated with a rise in the incidence of asthma attacks among adults with asthma three to five days after the pollution levels increased.

ENVIRONMENTAL IMPACTS OF PARTICULATE POLLUTION

Diesel particulate pollution from nonroad and stationary engines is a constituent of regional particulate problems leading to visibility impairment across the country. Fine particles in the lower atmosphere scatter and absorb light, obscuring scenic vistas such as those in national parks. Fine particles also play a major role in creating the "brown clouds" that shroud many western cities, particularly during the winter months.

OXIDES OF NITROGEN

Historically, NOx control strategies have been driven by the serious problem of ground-level ozone (smog), which generally occurs in the warm weather when NOx combines with volatile organic chemicals under certain atmospheric conditions to create ozone. The severity and frequency of asthma cases are exacerbated by ozone smog. A recent study suggests that exposure to elevated ozone concentrations can actually cause the onset of asthma. Ozone causes coughing, throat irritation and congestion in healthy adults. Millions of Americans live in areas that do not meet the health standard for ozone. Ozone pollution also damages plants, and costs the agriculture industry millions of dollars each year in decreased crop yields.

The dangers of ozone are reason enough to control NOx emissions from sources including diesel engines. But NOx pollution also contributes to the following serious public health and environmental problems that occur year round and require year-round control strategies in addition to those aimed at summer ozone: (1) formation of nitrate particles that contribute to harmful particulate pollution and obscure views; (2) acid deposition; and (3) eutrophication, or nutrient over-loads, in coastal waters that promotes unnatural algal looms that cloud the water and deprive submerged aquatic vegetation of the light necessary to grow.

SULFUR DIOXIDE

Nonroad diesel engines are a major source of sulfur dioxide, or SO_2, pollution by virtue of the high sulfur content in the diesel fuel used for nonroad applications. Just as NOx emissions convert in the atmosphere to nitrate, SO_2 pollution converts to sulfate, a fine particle implicated in the serious adverse health effects described earlier. Some studies have focused on the health effects of SO_2:

- A study in Seoul, South Korea found that stroke mortality increased in association with the concentrations of SO_2 and other pollutants.

- A study of children with asthma living in eight polluted U.S. cities found that SO_2 pollution was associated with an increase in morning asthma symptoms.

- In a 1985 study of an air pollution episode in Central Europe, 24-hour concentrations of total suspended particulates and SO_2 were associated with an increase in blood pressure.

Due to its transformation into sulfate particles, sulfur dioxide pollution also is one of the principal contributors to regional haze in national parks and brown clouds in western cities. It is also a major cause of acid deposition.

COMMERCIAL MARINE VESSELS

Diesel engines power the majority of the commercial shipping fleet used in both inland waterways of the United States and in oceangoing international shipping. The marine industry uses an estimated 10 percent of petroleum diesel fuel sold in the United States. Domestically, pollution from marine diesel engines has increased dramatically in the past decades.

Globally, diesel-powered marine engines also contribute a huge share of air pollution. Shipping traffic accounts for more than 14 percent of global sulfur emissions and more than 16 percent of global nitrogen emissions from petroleum use.

One of the central reasons shipping produces so much pollution is that marine diesel engines burn fuel far dirtier than any other diesel application. The U.S. inland shipping fleet runs on diesel fuel that can average 3,000–5,000 ppm [parts per million] sulfur, much like uncontrolled land-based nonroad engines. But the oceangoing fleet uses fuel containing 10 times as much sulfur, 30,000–45,000 ppm.

Oceangoing ships, associated harbor vessels such as tugboats, and dockside cargo loading equipment make ports air pollution hotspots. The two busiest U.S. ports, Los Angeles and Houston, are located in metropolitan areas with some of the country's worst air pollution. The American Lung Association has consistently awarded Los Angeles the title of smog capital of the year, and Houston recently ranked 5th. Los Angeles also

Dirtiest of the Dirty: Marine Residual Fuel

Among the dirty high sulfur fuels burned in diesel engines, none is dirtier than the residual fuel that powers large oceangoing ships. Residual fuel, also known as bunker fuel, is the tar-like product left behind after all the lighter petroleum fractions are refined from crude oil. Sulfur content in marine residual fuel can range as high as 45,000 ppm, or an astonishing 4.5 percent sulfur. EPA reports that, worldwide, residual fuel averages 27,000 ppm sulfur. This is more than 10 times the average sulfur level in the distillate diesel fuel used in smaller marine engines and land-based nonroad heavy equipment and nearly 2,000 times the 15 ppm level soon to be required for highway diesel fuels. The extraordinarily high sulfur content in residual fuel makes shipping one of the biggest sources of SOx emissions on the planet, despite the relatively small number of large ships in existence. And, as with other applications now powered by high-sulfur fuel, high sulfur levels make it impossible to apply most state-of-the-art emission control technologies to marine vessels.

EPA's newly issued standards for large marine engines do not place any limits on residual fuel sulfur content. Instead, EPA has left limiting this huge source of pollution up to the very uncertain prospects of an international treaty addressing shipping emissions. This treaty, known as MARPOL Annex VI, contains no general limit on residual fuel sulfur content, but does provide an opportunity to create SOx Emission Control Areas in which residual fuel cannot exceed 15,000 ppm sulfur. Given the uncertainty whether this treaty will ever be ratified, EPA has in effect indefinitely postponed addressing this critical source of pollution.

has some of the highest particulate concentrations in the country. In the Los Angeles basin, oceangoing ships, tugs and other commercial watercraft collectively account for 48 tons per day of smog-forming NOx emissions. This is comparable to the daily NOx pollution discharged by 1.5 million passenger cars and is nearly as much NOx pollution as the top 350 emitting industrial facilities in the basin.

Pollution from commercial marine engines is not restricted to coastal ports. Ninety percent of U.S. inland waterways are outside of port areas, and researchers have only now begun to understand that shipping emissions in these areas are a significant additional source of pollution. One study has estimated that commercial shipping traffic on inland rivers and the Great Lakes contributes 60 percent of the NOx emissions from all commercial shipping in the U.S., 33 percent of the total PM, and 48 percent of both HC and CO.

Emissions along inland waterways are concentrated in a compact area, much as highways concentrate emissions from cars and trucks. Thus, ship emissions along these waterways can be locally significant, even where they do not make up a significant portion of a regional or statewide inventory of emissions.

Researchers at Carnegie Mellon University have established that shipping traffic in the Pittsburgh area accounts for as much NOx emissions as a major metropolitan superhighway, and have concluded that "at least on an annual basis—waterborne transport can produce as much NOx as a region's freeways, even in large riverside cities (e.g. St. Louis, Nashville and New Orleans) where significant automobile commuting occurs."

Historically, marine engines have not been subject to any federal emissions controls. EPA promulgated standards for a limited number of commercial marine engines in 1999, but it has only begun to address this growing source of air pollution. EPA's December 1999 commercial marine vessel rule established limited standards for small- and medium-sized engines such as those used in ferries, tug boats and barges. But EPA's standards apply only to new engines, and will have no impact at all on existing engines that will be rebuilt for decades to come.

Notwithstanding their significant air pollution, EPA has taken only tentative steps toward controlling emissions from large oceangoing ships. In rules finalized in January 2003, EPA established a weak NOx standard that essentially codifies existing emissions levels for U.S.-flagged ships. The rule does not regulate foreign-flagged ships in U.S. territorial waters, which make up 95 percent of all calls to U.S. ports.

According to the American Association of Port Authorities, containerized shipping traffic in the United States doubled in

Cleaning up the Air Above Our Waters: Ports Begin to Reduce Air Pollution

Incentive programs have funded $22 million worth of air quality improvements at the Port of Los Angeles. Cranes that unload cargo have been converted to electric power, and smaller craft including tugboats and other harbor craft have been converted from old diesel engines to electric, compressed natural gas and newer, lower-polluting diesel engines. Operational changes are also yielding results. The Port's voluntary speed reduction program calls for ships in the port area to slow down to reduce NOx pollution. This simple step has lowered NOx emissions by an estimated two tons a day.

The Port of Houston has tested emulsified diesel fuel in yard tractors and cranes at the port and achieved NOx reductions of 25 percent, and PM reductions of 30 percent. It's possible that even greater reductions can be achieved. The Texas Waterways Operators Association has entered into an agreement with EPA, the Texas Commission on Environmental Quality and the Houston-Galveston Area Council to reduce NOx emissions from barges and other craft operated by the trade association's members. By replacing old engines, retrofitting existing engines, and reducing idling time, vessel operators expect to lower NOx emissions by 1.1 tons a day, which would be a significant step for the ozone-plagued Houston area.

the last decade, and is expected to double again in the next decade. The shipbuilding boom required to meet the growing demand for international shipping presents an opportunity to build cleaner, state-of-the-art marine engines that will begin to reverse the longstanding record of air pollution from shipping. Renewed federal leadership and innovative state and local measures are necessary to ensure that marine diesel engines do not remain a massive pollution source for decades to come.

How Has the Clean Air Act Helped Clean Our Air?

You can actually see the air quality problem in the brown haze that blankets many cities in the summer. If you live in one of these cities, it can become a visible menace. You may suffer from some sort of respiratory ailment. And it does not just harm the cities themselves. Pollution from cities hundreds of miles away makes its way to many of our national parks. In 2001, a letter from the Department of Interior to the U.S. Environmental Protection Agency (EPA) stated that "Visibility impairment is the most ubiquitous air pollution-related problem in our national parks and refuges . . . all areas monitored for visibility show frequent regional haze impairment."[1]

But the news is not all bad. The following sections are from a 2002 report called *Building on 30 Years of Clean Air Act Success: The Case for Reducing NOx Air Pollution*. Since the passage of the 1970 Clean Air Act, there have been many successes in the struggle to decrease air pollution. The levels of almost all of the pollutants listed in the National Ambient Air Quality Standards (NAAQS) have gone down.[2] Lead reduction is the biggest success, showing a 98% decrease. Carbon monoxide has decreased by 31%, and sulfur dioxide by 37%. Particulate matter has decreased by 71%. Despite these dramatic changes, the air is still polluted enough to cause health problems for plants, humans, and other animals.

The first selection explains the science and the problems of ground-level ozone. Although there have been successes in controlling ozone pollution, levels still remain high in many areas. The second section explains how nitrogen oxide (NOx) emissions add to the fine particle pollution in smog.

—The Editor

1. National Parks Conservation Association. *Code Red: America's Five Most Polluted National Parks*, 2002, p. 12. Available online at *http://www.npca .org/across_the_nation/visitor_experience/code_red/codered.pdf.*

2. Gutt, Elissa, et al. Environmental Defense. *Building on Thirty Years of Clean Air Success*. 2000. Available online at *http://www.environmentaldefense.org/documents/398_CAAReport.PDF.*

Building on Thirty Years of Clean Air Success
by Elissa Gutt et al.

GROUND-LEVEL OZONE AND SMOG NOx

Ozone, the main component of photochemical smog, forms in the lower atmosphere when NOx and various VOCs interact in the presence of sunlight, heat, and relatively stagnant air. Ozone was one of the first pollutants that the U.S. Environmental Protection Agency addressed under the Clean Air Act when it promulgated a health-based primary National Ambient Air Quality Standard for photochemical oxidants in 1971.

There has been significant progress on the ozone problem in much of the country. Across the United States, ozone concentrations measured over one hour declined by 20 percent between 1980 and 1999, and concentrations measured over eight hours declined by 12 percent.

Despite this progress, ozone still reaches unhealthy levels from Los Angeles, for a long time the home of the country's worst smog, to Houston, which is the new smog capital of the United States.

Ozone reaches unacceptably high levels across the country—in the interior West, in the South, around the Great Lakes, in the Ohio Valley, in the mid-Atlantic region, and in the Northeast.

We now understand that ozone is not just an urban problem. Ozone levels are increasing in many national parks. In 1999, the highest ambient ozone concentrations were found at suburban sites, and average one-hour ozone levels were greater in rural areas than in urban areas. It is clear that existing VOC reductions and NOx controls are not sufficient to lower these ozone levels and that in many areas NOx must be reduced to attain compliance with the ozone NAAQS and protect the population from ozone's harmful effects.

OZONE IS A SERIOUS THREAT TO PUBLIC HEALTH

Ground-level ozone causes chest pain, coughing, throat irritation, and congestion. People all over the country have become familiar with air quality alerts on high-ozone summer days. As ozone levels rise, more segments of the population are warned to avoid the outdoor air. Warnings are issued first to those with asthma or other respiratory impairments, then to the elderly and children, and finally, when ozone levels climb higher, even to healthy adults. In many areas, these warnings have become routine, and we tend to hear them without considering their implications. At the same time, many of us have experienced the symptoms of a dry throat and a burning sensation in the lungs when we are outside on high-ozone days.

In recognition of the steadily mounting evidence of the health and ecological harm caused by ozone, in 1997 the EPA revised the ozone NAAQS to include an eight-hour average standard of 0.08 parts per million (ppm), which will be phased in to replace the long-standing one-hour limit of 0.12 ppm. The EPA based this revision on "a significant body of information" demonstrating the presence of adverse health effects, particularly in children, at ozone concentrations permitted under the existing one-hour standard. . . . The new more protective eight-hour ozone standard, along with the revised particulate NAAQS, was remanded to EPA by a divided panel of the U.S. Court of Appeals for the D.C. Circuit. The revised ozone and particulate standards are now on appeal to the U.S. Supreme Court, where argument was heard on November 7, 2000.

We are cautioned to avoid physical activity outside when ozone levels are high because exertion makes us breathe harder and drives more burning ozone into our lungs. Children are particularly at risk because their lungs are not fully developed and so their airways are narrow, their respiration rate is higher in comparison to their size, and running and playing outside in the summer are the very activities that increase their exposure and risk. The elderly or ill are at risk because their respiratory function is frequently impaired and they cannot afford to lose any more of that function to ozone damage. Even healthy adults exercising outdoors or construction workers whose jobs require outdoor exertion can suffer a decrease in normal lung function on high-ozone days.

Although the sensation of burning that ozone causes is transient, its harmful effects may not go away when a cool front blows in fresh air. For example, children growing up in the Los Angeles basin have been found to suffer a 10 to 15 percent decrease in lung function compared with children growing up where the air is less polluted.

Recent public health research has revealed that respiratory illnesses in general, and asthma in particular, have become a health crisis in the United States. The incidence of asthma is growing dramatically. More than 14 million Americans now suffer from asthma, and the disease kills more than 5,000 people each year.

Even though we do not yet understand all the complex factors contributing to the asthma epidemic, it is becoming increasingly clear that high ozone levels trigger asthma attacks in those with the disease. Studies consistently show that asthma-related emergency room visits spike up on high-ozone days. Most recently, this ozone-asthma link was demonstrated in Sacramento, where scientists reviewed hospital records for the years 1992 through 1994 and found a 14 percent rise on high-ozone days in asthma-related emergency room visits and hospital admissions among low-income children and teenagers. In Sacramento County, where the air regularly exceeds state and federal ozone standards, about 5.5 percent of residents have asthma.

OZONE COSTS AGRICULTURE MILLIONS EACH YEAR

Plants are especially susceptible to ozone damage. Ozone enters plants through small pores in leaves, and once inside, it damages plants by interfering with their ability to produce and store food. This damage stunts growth and reproduction, reduces resistance to harsh weather, and, most visibly, causes leaves to brown, spot, or fall off.

Not only does ozone damage reduce the aesthetic value of the plants in yards and parks, but on a more easily quantifiable level, it takes a considerable toll on American agriculture. Ozone is "a significant stress factor in agricultural production in the United States." Studies of crop loss in individual states show the scale of ozone damage to agricultural crops.

In 1993, high ozone levels in California were estimated to have decreased the yields of cantaloupes, cotton, and grapes by 20 to 30 percent; alfalfa, dry beans, lemons, oranges, and onions by 10 to 15 percent; and corn, lettuce, rice, tomatoes, and wheat by 1 to 7 percent. In 1989, ozone was responsible for a 10 to 19 percent loss in California's cotton yield, a market worth more than $1 billion. Overall, ozone in California was responsible for estimated crop losses of $333 million in 1984 and $265 million in 1989. Similarly, it is estimated that Georgia's farm industry loses $250 million each year because of ozone-related crop damage. The EPA estimates that the cumulative net present-value agricultural benefits from Clean Air Act ozone abatement programs implemented over the period 1990 to 2010 "are on the order of $4 billion dollars."

TRANSFORMATION OF NOX INTO FINE PARTICLES

Fine particle pollution contributes to serious health impacts in communities across the country, haze in national parks and wilderness areas, and "brown clouds" in major western cities. In the atmosphere, NOx air pollution transforms into nitrates, which are small or fine particles. Ammonia, another form of nitrogen, also contributes to the formation of fine particles. Thus, as with acid deposition and eutrophication, both ammonia and NOx contribute to the effects of nitrogen—in this case, fine particles. Fine particles also result from the transformation of sulfur dioxide in the atmosphere and other pollutants that are directly emitted into the atmosphere as fine particles.

The importance of fine particle nitrates varies seasonally and regionally. The available data suggest that the role of nitrates has a strong seasonal dimension. Nitrates contribute more to fine particle concentrations during the nonsummer months.

These important nonsummer impacts further demonstrate the importance of year-round NOx reductions to effectively abate the suite of public health and environmental impacts associated with NOx air pollution.

NITRATES AND FINE PARTICLE HEALTH EFFECTS

Fine particles can penetrate deep into the lungs, and cause very serious health effects including hospitalization and death. Epidemiological studies have associated exposure to fine particles at pollution concentrations below the previous air quality health standards for particular matter with premature death, exacerbation of chronic disease, and increased hospital admissions. These findings led the U.S. Environmental Protection Agency in 1997 to establish more stringent national ambient air quality standards for fine particles. The EPA estimated that its new fine particle health standards would prevent about 15,000 premature deaths each year, "especially among the elderly and those with existing heart and lung disease."

Recently, a concerted research program has been undertaken to advance our understanding of fine particle health effects. Research by scientists in the United States and from around the world has confirmed the findings of the epidemiological studies underlying the EPA's new health standards—that fine particle concentrations are associated with very serious health effects, including premature death and hospitalization.

The major components of fine particles are "sulfate, strong acid, ammonium nitrate, organic compounds, trace elements (including metals), elemental carbon, and water." Some of these components are directly emitted as fine particles (primary particles), and others result from precursor pollutants that transform into fine particles in the atmosphere (secondary particles). "Sulfur dioxide (SO_2), nitrogen oxides (NOx), and certain organic compounds are major precursors of fine secondary [particulate matter]."

In 1999, the EPA and the states began deploying a new monitoring network to evaluate fine particulate air pollution concentrations. When fully deployed, the fine particle monitoring network will consist of about 1,700 monitors at more than 1,100 sites across the country. Currently, the only existing long-term fine particle monitoring network is that managed jointly by the National Park Service, the EPA, and the states to assess visibility conditions in national parks (the Interagency

Monitoring of Protected Visual Environments, or IMPROVE). Most of the IMPROVE monitors are located in rural areas, in relatively remote national parks. Consequently, we currently have limited information about fine particle concentrations and the contribution of the different pollutant components to fine particle levels, especially in urban areas. One of the IMPROVE monitors, however, is located in Washington, D.C., and several others are sufficiently close to metropolitan areas to indicate the contribution of nitrates to fine particle concentrations in urban areas with elevated fine particle concentrations.

The IMPROVE data indicate that during the winter in Washington, D.C., nitrates comprise about 17 percent of the fine particle mass budget, with both sulfates and organics adding about 33 percent each. In the San Gorgonio Wilderness east of Los Angeles, nitrates are the largest contributor to fine particle mass in the spring, fall, and winter, likely reflecting the high levels of NOx air pollution in nearby Los Angeles. A monitor at the Lone Peak Wilderness Area north of the sprawling metropolitan area along the Wasatch Range in northern Utah indicates that nitrates are the predominant contributor to fine particle mass in the winter, making up about one-third of fine particle pollution concentrations.

The IMPROVE data indicate that nitrates are a significant percentage of fine particles in areas with pollution levels near or above the health-based ambient air quality standards. This is the case in both the eastern and western United States. The IMPROVE data also show that in a number of areas across the country, sulfates and organics are the main contributors to fine particle pollution. Sulfate concentrations, especially in the East, have a key role in contributing to fine particle concentrations, reflecting the high level of sulfur dioxide pollution from the electric utility sector.

NITRATES AND HAZE AIR: POLLUTION IN NATIONAL PARKS
The fine particles that have deleterious health impacts also impair visibility. Our views of landscapes, geological formations, and breathtaking vistas are impeded by fine particles

that scatter and absorb light. According to the National Academy of Science's National Research Council, "[s]cenic vistas in most U.S. parklands are often diminished by haze that reduces contrast, washes out colors and renders distant landscape features indistinct or invisible."

The ability to enjoy scenic vistas is integral to the quality of visitors' experience. Indeed, the National Park Service has found that visitors rank clean, clear air among the most important features of national parks. Visitors feel so strongly about clean air in the national parks that they are willing to pay additional entrance fees to obtain better visual air quality during their park visit, and households are willing to pay to preserve visual air quality in premier national parks, regardless of whether they intend to visit the area.

Since the 1977 Clean Air Act Amendments were enacted, the statute has had a program specifically designed to protect the scenic vistas in our premier national parks and wilderness areas. Congress adopted the visibility program to protect the "intrinsic beauty and historical and archeological treasures" of certain federal lands, observing that "areas such as the Grand Canyon and Yellowstone Park are areas of breathtaking panorama; millions of tourists each year are attracted to enjoy the scenic vistas." To guide the administration of the Clean Air Act's visibility protection program, Congress declared and codified a national visibility goal: "the prevention of any future, and the remedying of any existing, impairment of visibility in [premier national parks and wilderness areas] which impairment results from manmade air pollution."

Policy solutions to curb the haze in national parks generally have focused on reducing SO_2, because in many national parks the resulting sulfate particles are the principal component of visibility impairing pollution. However, in a number of national parks, nitrate particles also are a significant contributor to the haze pollution and, in some instances, are the main source of visibility-impairing fine particles. Comprehensive strategies to clear the manmade haze in our national parks thus must include not only reductions in SO_2 but also cuts in the other pollutants, such as NOx and ammonia, that significantly contribute to haze air pollution in some areas.

Nitrates impair visibility in national parks more during the nonsummer seasons, and in a number of national parks the nitrate component of wintertime visibility impairment is considerable. In nationally protected areas across California, including the Redwood National Park, Point Reyes National Seashore, Yosemite National Park, and San Gorgonio Wilderness, nitrates were the predominant cause of poor visibility. Nitrates also had a substantial role in the visibility impairment monitored in the central northern part of the country, including the Badlands in South Dakota and the Boundary Waters in northern Minnesota.

In addition, nitrates make up nearly one-fifth of the impairment problem during the winter at Rocky Mountain National Park. Further, nitrates can be the largest source of fine particles at Rocky Mountain National Park on those winter days with the worst visibility. This is not entirely surprising in light of the high wintertime concentrations of nitrates that contribute to Denver's brown clouds. The high wintertime levels of nitrates at Rocky Mountain National Park may be due to "intrusions of haze from Denver."

NITRATES AND "BROWN CLOUDS" IN WESTERN CITIES

Blue skies, rugged mountains, and scenic vistas have spurred population and economic growth in cities across the interior western United States. But this growth has led to a variety of air pollution problems.

A number of western cities—including Denver, Phoenix, and Salt Lake City—now experience "brown clouds" that enshroud these cities in a dirty pollution haze, usually during the winter months.

Measures to clean up these brown clouds will help ensure that the crisp blue skies and clear vistas continue to attract new residents and visitors. In many of these cities, particulate nitrates are a significant contributor to the problem, along with other fine particle components.

The "Brown Cloud" in the Valley of the Sun

Phoenix, Arizona, enjoys some of this country's most spectacular wintertime weather. The mild winters and bright, sunny days have inspired the city's moniker: the Valley of the Sun. The climate also has inspired flocks of wintertime visitors—so-called snowbirds—who leave cold climates to spend their winters in Phoenix, as well as an influx of permanent residents.

The resulting tourist-related and industrial growth has led to a tremendous economic expansion but at the same time has exacerbated the brown cloud pollution problem.

NOx emissions contribute about 18 percent to the worst visibility impairment during the fall and winter in Phoenix, which is second to direct particulates resulting from fuel combustion. In response to public concern about the brown clouds and analysis indicating that Phoenix's brown clouds were intensifying, the governor of Arizona recently issued an executive order convening a brown cloud summit to "examine strategies to improve visibility in the Valley of the Sun."

Denver's "Brown Cloud"

On clear winter days, the views from Denver to the Rocky Mountains are sweeping and inspiring. On winter days with high levels of fine particle pollution, the brown cloud that settles over Denver obscures these views and makes Denver indistinguishable from many other large, polluted metropolitan areas.

In 1995, Colorado policymakers called for an air pollution assessment of the pollutants and sources that form Denver's brown cloud. The study found that particulate ammonium nitrate, formed from emissions of NOx and ammonia, contributed about 25 percent of the fine particles that comprise Denver's wintertime brown cloud and that gasoline vehicle exhaust added about 28 percent of the fine particles. Thus, particulate ammonium nitrate was the second most prevalent species contributing to wintertime fine particle levels.

The sources of NOx that contribute to nitrate concentrations include gasoline- and dieselpowered cars and trucks, nonroad diesel engines, coal-fired power plants, natural gas combustion, and industrial sources. The ammonium is largely due to pollution from agricultural sources. Agricultural sources are responsible for most of the ammonia emissions in the area, producing about 97 tons, or 85 percent of the estimated 114 tons of the daily ammonia emissions.

Colorado has adopted a visibility standard for the metropolitan Denver area to help assess progress in cleaning up the brown cloud. While the number of days the standard has been exceeded since 1990 has varied, the record over the past decade suggests that there has not been significant progress in cleaning up the brown cloud. Most recently, the standard was exceeded on about 70 days in the winters of 1997–98 and 1998–99. This frequency is significantly higher than the 55 days on which the standard was exceeded during the initial monitoring in the winter of 1990–91.

Is Air Pollution a New Problem?

It is not only people who pollute the air. There are natural sources of air quality problems. Active volcanoes spew huge amounts of dust into the atmosphere. Forest fires and dust storms are other natural polluters. On a less dramatic scale, plants give off spores, pollen, and even some volatile hydrocarbons. But unless you live near an active volcano, natural pollution does not pose serious human health problems for several reasons. Usually, the major sources of natural pollution are found far from where people live. In addition, the level of contaminants are low, and major sources (like volcanoes, forest fires, and dust storms) do not last very long.[1]

Historically, air quality troubles began when large groups of people started to burn things. At first, it was wood smoke that clouded the air, then it was coal smoke. Today, fuel-burning industries constantly add more pollutants to the air.

The following sections from Richard P. Turco's 2002 book *Earth Under Siege: From Air Pollution to Global Change* present a brief history of humans and air pollution. Originally, the term *smog* was used to describe the mixture of fog and smoke from coal-burning plants in England. The gray smoke was called "London smog." More recently, a new type of smog is found in cities around the world. It is a photochemical smog born of sun-heated car and truck emissions. This smog shows up as the mucky brown haze that surrounds many cities, especially in the summer.

Common sense suggests that we learn what we can from our past mistakes. Toward that end, these sections offer some history of air quality problems, to show us how to avoid polluting the air as we move into the future.

—The Editor

1. Goddish, Thad. *Air Pollution*. Boca Raton: Lewis Publishers, 2004, pp. 23–24.

Earth Under Siege:
From Air Pollution to Global Change
by Richard P. Turco

SMOG: THE URBAN SYNDROME

Cities in developing and developed countries alike are being smothered by soupy clouds of tainted air. As civilization continues its industrial development and population expands, the production and emission of unhealthful compounds into the environment accelerate. In some places, restraints are placed on polluters. But there are too many places where the laws discouraging pollution are weak or the enforcement of laws is halfhearted. The serious polluters escape to these places, or they relieve themselves of polluting waste under cover of darkness. When economic factors are considered, pollution control takes a back seat to business development and jobs. Millions of subsistence farmers leave the land and flock to cities in search of menial work in hazardous air. Facilities and infrastructure lag behind the needs of these masses. Manufacturing is done under primitive conditions in which pollution abatement is not a priority. Developed countries export the most hazardous industries to the Third World. The eastern European nations of the former Soviet bloc struggle with inefficient, highly polluting facilities to earn a marginal living. Sophisticated technological fabrication introduces new pollutants with unknown long-term effects. Minute quantities of dangerous compounds are detected in the atmosphere, water, and soil.

Urban air pollution, or *smog,* is not a recent problem or the most serious environmental threat that we now face. However, pollution in cities affects hundreds of millions of people, leading to illness and general malaise. Santiago, Chile, for example, is one of the smoggiest cities on Earth. Santiago has a coastal mountain topography extremely well suited to trapping smog. Almost two decades of communist rule with disregard for controls on development and pollution led to the predicament. The largest polluters in the city are privately owned buses, once encouraged by the government in order to provide cheap transportation. The noxious diesel fumes emitted by the unregulated fleet of 11,000 buses has created smog so dangerous that children are often kept indoors to prevent their breathing the air. Only recently have new policies been adopted to remove the oldest buses from service, but Santiago still remains shrouded in a pall of bad air.

THE HISTORY OF SMOG

When small groups of humans roamed the Earth as hunter-gatherers, there were no cities and little air pollution. To be sure, there were many natural sources of gases and particles that would be considered pollutants if emitted from industrial facilities today. Organic vapors from trees, smoke from vegetation fires, sulfur fumes from volcanic vents, and fetid vapors from swamplands all cause natural air pollution. Unless one is unlucky enough to live next to a stinking swamp or an erupting volcano, however, natural pollution is generally a minor bother. Rather, the pollution we are concerned with nowadays, which evolved with human industry and technology, is dense and dangerous and, for millions of us, unavoidable.

When agriculture developed into a major human enterprise, villages and towns sprang up as permanent residences for most people. As these embryonic cities grew, the waste from human activities and its disposal became a serious problem. Even when the Spanish explorer Juan Rodriguez Cabrillo first set anchor in October 1542 in San Pedro bay near the present site of Los Angeles, the air was clouded by the smoke from Native American campfires. He named the harbor the Bay of Smoke. Although the ravaging effects of air pollution were known well before that time, the problem had not yet been systematically described. In ancient Rome, the blackening of buildings by smoke from wood fires was noted in passing by the chronicler Horace (b. 65 B.C.). Seneca (b. 3 B.C.), Nero's tutor, noticed that his health improved markedly once he left the "oppressive fumes and culinary odors" of Rome.

Coal may have been used as a fuel by the Chinese as early as 1000 B.C. It was identified and used in Europe by at least around

A.D. 1200. The early natural scientist Moses Maimonides wrote of the poor quality of air in cities even in the twelfth century:

> Comparing the air of cities to the air of deserts and arid lands is like comparing waters that are befouled and turbid to waters that are fine and pure. In the city, because of the height of its buildings, the narrowness of its streets, and all that pours forth from its inhabitants and their superfluities . . . the air becomes stagnant, turbid, thick, misty and foggy . . . If there is no choice in this matter, for we have grown up in cities and have become accustomed to them, you should select from the cities one of open horizons . . . endeavor at least to dwell at the outskirts of the city.
>
>   . . . [I]f the air is altered ever so slightly, the state of the Psychic Spirit will be altered perceptibly. Therefore you find many men in whom you can notice defects in the actions of the psyche with the spoilage of the air, namely, that they develop dullness of understanding, failure of intelligence and defect of memory.[1]

The extensive burning of coal did not begin until the early eighteenth century, with the discovery of a process for making coke (another solid form of coal) and coal gas.[2] Nevertheless, Eleanor, queen consort of King Henry III of England, reportedly complained around 1250 about the pollution created by the burning of coal. King Edward I, the son of Henry III, later (ca. 1300) issued a proclamation against the use of coal (presumably around the palace where it upset his wife, another Eleanor): "Be it known to all within the sound of my voice, whosoever shall be found guilty of burning coal shall suffer the loss of his head."

The use of coal accelerated despite Edward's dissatisfaction with its side effects. In his classic work *Fumifugium, or The Inconvenience of the Air and Smoke of London Dissipated* (1661), John Evelyn described the air quality in seventeenth-century English cities:

It is this horrid smoake which obscures our church and makes our palaces look old, which fouls our cloth and corrupts the waters, so as the very rain, and refreshing dews which fall in the several seasons, precipitate to impure vapour, which, with its black and tenacious quality, spots and contaminates whatever is exposed to it.

. . . [I]t is evident to every one who looks on the yearly bill of mortality, that near half the children that are born and bred in London die under two years of age (a child born in a country village has an even chance of living near forty years). Some have attributed this amazing destruction to luxury and the abuse of spiritous liquors. These, no doubt, are powerful assistants; but the constant and unremitting poison is communicated by the foul air, which, as the town still grows larger, has made regular and steady advances in its fatal influence.[3]

In fact, the use of coal was the backbone of the *Industrial Revolution*, which began in mid-eighteenth-century England. The rapid expansion of manufacturing based on steam energy generated from coal combustion was the early hallmark of the revolution. *Steam* is nothing more than heated water vapor. James Watt invented an engine that could be driven by steam, which expanded into piston chambers, much as burning gasoline expands in the cylinders of an internal-combustion engine. Naturally, the levels and extent of the accompanying air pollution also rose dramatically.

AIR POLLUTION AND POETS

In the social and business culture of the Industrial Revolution, which fostered sweat shops and child exploitation, the environmental effects of industrial pollutants could be completely disregarded. Degeneration in the quality of life was noticeable to all, however. Over the years, writers and poets have captured the feeling of grayness and depression during this period. William Shakespeare himself serendipitously expressed an early opinion on the state of the atmosphere and sky in English cities before the Industrial Revolution: "This most excellent canopy, the air, look you, this brave o'erhanging firmament, this majestical roof fretted with golden fire, why, it appears no other thing to me than a foul and pestilent congregation of vapours" (*Hamlet*, Act 2).

The great poet Percy Bysshe Shelley,[4] who lived during the early period of the Industrial Revolution, was very explicit in his description of London at the time:

> *Hell is a city much like London—*
> *a populous and smoky city.*
>> *Peter Bell the Third*, Part III, Stanza I

Similarly, William Morris (1834–1896) contrasted two visions of London:

> *Forget six counties overhung with smoke,*
> *Forget the snorting steam and piston stroke,*
> *Forget the spreading of the hideous town;*
> *Think rather of the pack-horse on the down,*
> *And dream of London, small, and white, and clean.*
>> Prologue to *The Earthly Paradise*

By the mid-nineteenth century, the Industrial Revolution had spread to the rest of Europe and the United States. The building of machines and the introduction of new technologies and materials picked up speed, and the degradation of the environment did as well. Listen to James B. Dollard (1872–1936) from the countryside in Scotland:

> *I'm sick o' New York City an' the roarin' o' the thrains*
>> *That rowl above the blessed roofs an' undernaith the dhrains;*
> *Wid dust an' smoke an' divilmint I'm moidhered head an' brains,*
> *An' I'm thinkin' o' the skies of ould Kilkinny!*
>> *"Ould Kilkinny!"*

But every dark cloud has its silver lining. So William Henry Davies (1871–1940) wryly noted the pleasing side effects of air and water pollution:

> What glorious sunsets have their birth
>   In cities fouled by smoke!
> This tree—whose roots' are in a drain—
>   Becomes the greenest oak!
>
> "Love's Rivals"

One of the leading thinkers of the twentieth century, Buckminster Fuller,[5] grasped the true nature of modern-day pollution of the environment by recognizing that "pollution is nothing but resources we're not harvesting."

LONDON SMOG

Beginning in the mid-nineteenth century and extending through the first half of the twentieth, the major cities of Europe and the United States experienced episodes of choking air pollution associated with the burning of coal to generate heat and energy. Thousands of people died as a result of exposure to these toxic palls. The most serious event occurred in London in December 1952. The British Isles were capped by a large-scale temperature inversion and blanketed in dense fog. For five days, from December 5 to 9, air pollutants accumulated in the Thames River valley in stagnant air. About 4000 excess deaths were attributed to the inhalation of smoke, sulfurous particles, and soot mixed with fog.[6] Most of the victims suffered respiratory and heart failure. Thousands of others, especially asthmatics and people with bronchitis and other respiratory ailments, were left gasping for oxygen. The very young and very old were most vulnerable.

London had experienced killer pollution events in December 1873 (1150 dead), January 1880, February 1882, and December 1891. All these tragedies had one thing in common: The deadly conditions were precipitated by the combination of a stagnant fog and the smoky emissions of coal. In 1905, Harold Antoine des Voeux, a medical doctor, first used the term *smog*, which combines the words smoke and fog to describe the dark palls he observed hanging over many British towns. The term became popular when he published a report in 1911 on a killer smog episode in Glasgow, Scotland, which in 1909 killed 1063 residents. We will refer to this type of urban pollution as *London smog*.

As the Industrial Revolution spread, so did the killer London smog episodes. A notable nasty pall developed in the Meuse Valley, Belgium, from December 1 to 5, 1930, resulting in 63 deaths and general misery for many others. The first major smog episode in the United States occurred in Donora, Pennsylvania, from October 26 to 31, 1948. In this incident, 20 excess deaths were recorded, and 43 percent of the population fell ill. Poza Rica, Mexico, suffered 22 deaths in November 1950 when hydrogen sulfide gas escaped from a natural gas facility under temperature inversion conditions similar to those that cause London smog. New York City has had a number of serious smog events, including one from November 12 to 22, 1953, that engulfed the entire metropolitan area with less severe but very widespread effects. Subsequently, from November 24 to 30, people were recorded as victims of smog in the New York area. In all these incidents, thousands of persons became seriously ill but recovered. The long-term damage to their health and the health of other millions exposed to the pollution may never be assessed.

The typical London smog results from the accumulation of smoke from coal burning. This smoke has a high sulfur content and leads to the production of high concentrations of sulfuric acid in fog droplets. These acidic particles, along with high densities of smoke, inhibit the normal functioning of the lungs. The symptoms include chest constriction, difficulty in breathing, headache, nausea, vomiting, and eye, ear, nose, and throat irritation. . . . The episode corresponded to an extended period of fog, during which the death rate (excess deaths per day) soared. Effective legislation to control smoke emissions throughout Great Britain was passed in 1956.

LOS ANGELES SMOG

Today the term *smog* is used to describe another type of air pollution experienced in many cities around the world. This smog is not derived mainly from smoke and fog; rather, the emissions of automobiles and other vehicles are the primary cause. This different kind of smog forms when the meteorological conditions are right—that is, in stagnant air capped by a strong temperature inversion and illuminated by plenty of sunlight. Cities like Los Angeles are ideal for the production of this type of smog. In addition to the temperature inversions and sunlight, Los Angeles is overrun with cars. The mixture of ingredients that are emitted from automobiles reacts in the presence of sunlight to create high ozone concentrations and brown haze. Accordingly, this pollution is known as *photochemical smog*. We will also refer to it as *Los Angeles smog*, since this was the place where such smog was originally identified.

Los Angeles developed an early love affair with the automobile, installing the country's first automatic traffic signals in 1922. The potential problems associated with photochemical smog began to appear there in the early 1940s. The haze already common in the Los Angeles basin started to thicken. Catalina Island, off the coast, and the majestic San Bernardino and San Gabriel Mountains disappeared from view more frequently. Agricultural crops began to show signs of damage, particularly the bronzing of foliage, which was most noticeable on parsley. The rubber in car tires and tubing showed premature aging and cracking. According to newspaper accounts, citizens were weeping, sneezing, coughing, and complaining. By 1947, the problem was considered serious enough to set up the first Los Angeles Air Pollution Control District.

Scientific studies revealed that some component of the polluted air was damaging crops. Although the compound could not be positively identified, it was proved not to be sulfur dioxide (the key ingredient of London smog). In 1951, Arie Haagen-Smit, working in Los Angeles, showed through laboratory simulations that mixtures of hydrocarbon vapors and ozone smell the same as the smog does, and cause leaves to bronze in the same way. He suggested in 1952 that ozone actually forms in air containing hydrocarbons and nitrogen oxides. Further experiments carried out soon thereafter confirmed the Haagen-Smit theory of photochemical smog formation.[7]

In fact, ozone had been detected in urban air during the second half of the nineteenth century. Christian Schonbein, who first identified ozone as a form of oxygen, also designed a crude ozone detector. The instrument, called an *ozonometer*, consisted of a specially treated paper strip that reacted with ozone, causing its color to bleach out. The amount of fading determined the ozone concentration, which Schonbein calibrated in his laboratory. These novel ozonometers soon appeared all around Europe. One of them, situated near Paris, provided a 30-year record of ozone concentrations between 1876 and 1907. During this period, the average ozone concentration in that locale was about 10 parts per billion by volume (ppbv). That amount is comparable to the abundance of ozone in "clean," or natural, tropospheric air today. By the mid-twentieth century, average ozone concentrations had increased in rural Europe to 20 to 30 ppbv, still below the average concentrations of 100 to 200 ppbv measured in heavily polluted air. Of course, Schonbein never figured out how ozone was actually formed in the atmosphere. More sophisticated analytical instruments and some imaginative thinking on the part of Haagen-Smit were needed.

The recognition of the cause of smog in Los Angeles did not stem the tide of bad air. After World War II, in fact, with population and industry in the area booming, air pollution soared. By 1958, Los Angeles was experiencing 219 days with stage-1 smog alerts. At such concentrations, air pollution is considered to be hazardous to health.

NOTES

1. Moses Maimonides (b. 1135 in Cordoba, Spain; d. 1204 in Egypt) was a Jewish philosopher, jurist, and physician. He wrote a classic code of Jewish law, *The Guide of the Perplexed*. He contributed to both science and religion in his lifetime. Quotation is from V. Goodhill,

"Maimonides—Modern Medical Relevance," *Transactions of the American Academy of Ophthalmology.*

2. When coal, especially bituminous coal, is heated in the absence of air, the volatile components are given off as vapors, consisting of methane, hydrogen, and other "coal gases." The residual solid, coke, is almost pure carbon. Coal gas can be employed as a fuel and in the past was used for illumination.

3. John Evelyn (1620–1706) was a country gentleman who chronicled life in England during his lifetime. He wrote 30 books, treatises, and discourses on various subjects, including forest silviculture and stamps, and kept a diary, later published, covering more than 50 years.

4. Percy Shelley's wife, Mary Wollstonecraft Shelley, wrote the famous novel *Frankenstein* in 1816.

5. Richard Buckminster Fuller (1895–1983) never formally completed college although he was nonetheless accomplished at architecture, engineering, philosophy, cartography, and poetry. He invented the geodesic dome, the only known structural design that can be scaled to any size without collapsing. Following the death of his daughter at the age of 4 from influenza and polio, he dedicated his remaining years to designing environmentally safe and efficient technologies and industries. He engineered the first streamlined car with omni-directional steering (you could park sideways) and fully surrounding bumpers (like a "bumper car"), the Dymaxion. His 1943 design promised 40 to 50 miles per gallon of gasoline, but the car was never produced. Fuller felt that a "comprehensive and anticipatory" approach to design, demonstrated by the Dymaxion, could solve the world's problems of housing, hunger, transportation, and pollution.

6. The term *excess deaths* refers to the additional number of fatalities counted above the number expected under otherwise normal conditions. The latter number is determined using statistics on mortality rates during normal periods.

7. Arie Haagen-Smit is considered to be the "father of photochemical smog." He not only discovered the cause of the Los Angeles air pollution, but also campaigned vigorously to see that tough rules to control smog were put into effect. Although strongly resisted by special interests and hindered by public apathy, he eventually prevailed.

Is Acid Rain Still an Environmental Issue?

What would happen if you dumped lemon juice into a tank filled with plants and fish? You really do not have to do the experiment to know that everything would die because the lemon juice is so acidic. Knowing this, you can see that when certain pollutants mix in rainwater and make it acidic, it is not surprising that fish in lakes and even entire forests can die. This phenomenon is called acid rain, or acid precipitation, and it is not a new problem. In 1872, Robert A. Smith published a book called *Air and Rain: The Beginnings of a Chemical Climatology*. In it, he first introduced the term *acid rain*. He had been studying the sulfur and ammonia concentrations in rain around coal plants in England. In the 1950s, acid rain was recognized as a growing problem in Europe. In the 1980s, the United States and Canada began to coordinate efforts to reduce acid rain.[1]

Acid rain forms when emissions from industry, especially coal-fired plants and internal combustion engines, produce oxides of sulfur and nitrogen. In 1990, the Clean Air Act was amended to try to curb the pollutants that cause acid rain. But as new research continues to show, the problem has not been solved. In July 2004, the EPA agreed to a proposed legal settlement to strengthen the requirement for industrial engines and power plants. "EPA's own estimates show that public health and other benefits would outweigh the cost by a 20 to 1 ratio," says Mark McCloud of Environmental Defense.[2]

The following report by the Clean Air Task Force outlines the problems that still exist today and points to the directions we need to take in solving the problem. The Clean Air Task Force is a nonprofit organization dedicated to restoring clean air through scientific research, public education, and legal advocacy.

—The Editor

1. Turco, Richard P. *Earth Under Siege: From Air Pollution to Global Change*. New York: Oxford University Press, 2002, p. 260.

2. Environmental Defense. News Release. August 2004. *28 Eastern States Would Benefit from Stronger EPA Power Plant Rule*. Available online at *http://www.environmentaldefense.org/pressrelease.cfm?ContentID=3904*.

Unfinished Business:
Why the Acid Rain Problem Is Not Solved
from the Clean Air Task Force

EXECUTIVE SUMMARY

Over a decade ago, widespread damage from acid rain to forests and waters prompted government action to reduce the threat. In North America, the US Clean Air Act, the Eastern Acid Rain Program in Canada and cooperative agreements between the US and Canada led to significant declines in emissions of sulfur dioxide across large regions of North America. With the passage of the 1990 Clean Air Act Amendments, many hoped that its 50 percent reduction in sulfur dioxide (SO_2) emissions—a cut of 10 million tons—and a reduction of two million tons of oxides of nitrogen (NOx) from the utility industry would be enough to support swift recovery of ecosystems damaged by acid rain.

There is no question that legislation to date has made a positive difference. Reductions in emissions of acidic compounds have been followed by reductions in deposition, and chemical improvements in some soils and waters have been documented. But despite the emissions reductions, the problem of devastated forests, lakes, streams and ecosystems due to acid rain has not been solved. A growing body of evidence shows that without significant additional cuts in acid rain-forming emissions many of the problems associated with acid rain will persist for many, many decades.

The reason for this is that even with current and anticipated reductions, emissions are still much higher than preindustrial levels. Over the past 150 years, millions and millions of tons of acidic compounds have fallen to the earth's surface. This long history of acid burden means that ecosystems throughout North America remain damaged and at risk. Since passage of the Clean Air Act Amendments of 1990, scientists better understand ways in which acid rain alters ecosystems. We now know that if we are to reverse the chemical effects of acid rain accrued over a century and a half, we must further reduce emissions of acidic compounds. This report documents that: Acid rain is still a major problem.

Even after the cuts required by the current Acid Rain law have taken place:

- Atlantic salmon populations will continue to decline in Nova Scotia.

- Nearly 100,000 Canadian lakes will be damaged.

- Acid-sensitive streams in New York's Catskill and Adirondack Mountains will be too acidic to support a diversity of aquatic life.

- Thirty percent of brook trout streams in Virginia will not be able to support brook trout.

- Reductions in fish diversity will persist in northwest Pennsylvania.

- Declining vigor of red spruce and sugar maple will likely occur in other tree species as well.

Greater emission reductions will be needed to support recovery Analyses conducted at sites in New Hampshire, New York and Virginia indicate that only with deeper cuts in emissions of acid compounds—up to 80 percent beyond Clean Air Act Amendments of 1990—will biological recovery be able to begin by mid century in acidsensitive areas of North America.

Economical technology is available to achieve the level of deep reductions needed. The needed reductions can come today from a mixture of energy options including expansion of the nation's use of energy efficiency, clean renewables, cleaner fuels and pollution control equipment.

ACID RAIN MEANS STRONG ACIDS HIT THE EARTH'S SURFACE

In 1999 there were many parts of the US where the annual, average pH of wet precipitation was below 4.5, ten times more acidic than normal precipitation. For the average pH to be in the low fours, some rain events have to be even more acidic than that.

Since 1999, NADP [National Atmospheric Deposition Program] data show weekly pH values of wet deposition (rain or snow) have fallen below four—the acidity of tomato and orange juice—in many states, including Ohio, Indiana, Illinois, Pennsylvania, Maryland, Virginia, West Virginia, Kentucky, North Carolina, Tennessee, Georgia, South Carolina, New York, Vermont, Maine, Connecticut, New Hampshire, Massachusetts, New Jersey.

ACID RAIN AFFECTS SOIL CHEMISTRY

Soils contain many substances including aluminum, calcium and magnesium. When acid compounds enter soils, there is some plant uptake, and some of the compounds move into ground and surface waters. Still others stick to soil particles and, in doing so, replace calcium and magnesium, which dissolve and enter ground and surface waters during rains and snowmelt. Acids also mobilize aluminum—which is abundant in soils in a harmless, organic form. Once released, however, the organic form is converted to inorganic aluminum, which is toxic to living organisms.

Soil depth also makes a difference. For instance, southern soils are generally deeper than northern ones and create a more effective buffer against acid damage. Some southern soils are just now becoming acid saturated and are no longer able to retain acids. As a result, acid levels in waters in the southeast US are now increasing.

SITE SENSITIVITY DEPENDS ON DEPOSITION AND GEOLOGY

How a site is affected by acid rain depends on the levels and history of acid deposition, combined with its "sensitivity" or ability to neutralize acidic inputs. The most sensitive sites lack soils that are able to neutralize the acidic inputs. Soils with good neutralizing capacity are rich in base cations, mostly calcium and magnesium.

When sulfur dioxide and nitrogen oxides are emitted into the atmosphere, they come into contact with water where they are chemically converted to acidic compounds of sulfates and nitrates. These strong acids are deposited onto the earth's surface as rain, snow and fog and through dry deposition. While acidic deposition is the more accurate term, acid rain is used more commonly.

Acid rain comes from the burning of fossil fuels. Sulfur, an impurity found in coal and oil (and in trace amounts in natural gas), is released when fossil fuels are burned, largely for electricity production and industrial processes. Oxides of nitrogen are released during burning of all fossil fuels, including gasoline and diesel fuel, when the nitrogen in the fuel and atmosphere reacts with oxygen.

Most sensitive sites receive deposition that exceeds the soils' ability to neutralize acids. Many of these occur downwind of emission sources, often in mountains where soils are thin as well as poorly buffered, i.e. low in basic compounds. These high elevation sites are also more vulnerable because mountain fog is frequently more acidic than rain.

In the US, the areas that combine acid rain and little neutralizing capacity are located at high elevations east of the Mississippi. These are: the southern Blue Ridge Mountains of eastern Tennessee, western North Carolina and northern Georgia; the mid Appalachian Region of eastern West Virginia, western Virginia and central Pennsylvania; New York's Catskill and Adirondack Mountains; the Green Mountains of Vermont; and the White Mountains of New Hampshire.

In some cases, extremely poor buffering capacity means that the site cannot even neutralize small increases in acidity. This condition exists along the coast of Nova Scotia.

Many sites also display moderate sensitivity, such as cation—poor sandstone soils in southern Indiana and Ohio and calcium-depleted soils in Georgia, South Carolina and North Carolina.

Some of the least acid sensitive sites in the country are located in parts of the Great Plains where acidic deposition is relatively low, and there are deep, calcium-rich soils. In these well-buffered soils, large stores of calcium and magnesium can handle acid inputs with little apparent impact for decades or sometimes even centuries. However, even in these cases,

150 years of acidic deposition means retention of acidic compounds well above historic backgrounds.

AQUATIC SYSTEMS

Aquatic species are affected both by episodic and chronic acidification. The more acidic a lake or stream becomes, the fewer species it can support. Plankton and invertebrates are among the first to die from acidification, and when the pH of a lake drops below 5, more than 75 percent of its fish species disappear.

Some aquatic species can handle acidic conditions. Many others, though—mayflies, some crayfish, lake trout—decline as pH decreases. The result is that biological communities in acidified lakes have fewer species (less biodiversity) than water bodies that are not acidic. This reduction in biodiversity matters.

As diversity is diminished, ecosystems become less stable and productive. When diversity is lost, the quality of life for all is diminished, and there is a greater risk that critical parts of the cycle of life will fail. . . . .

Chronic Acidification, Episodic Acidification and Acid Neutralizing Capacity (ANC)

Acid Neutralizing Capacity (ANC) is the term used to describe the ability of a water body to counteract or neutralize acid deposition. ANC is measured in microequivalents per liter (μeq/L). When lakes and streams have an ANC that is always below 0 μeq/L, there is no ability to neutralize acidic inputs, and the waterbody is considered chronically acidic. More commonly, there are short bursts of high acidity that come after snowmelt or heavy rains when the ANC falls below 0 for hours or even weeks. These surface waters are described as being episodically acidic. This "acid shock" usually occurs in the spring often before the growing season but can occur in the fall as well. Surface waters with ANC values greater than 50 are considered insensitive to inputs of acidic deposition. Although ANC values above 50 are common in North America, acid sensitive sites often have ANC values below 0.

MAJOR PROBLEMS TO AQUATIC LIFE
FROM ACIDIC RAIN INCLUDE:
Loss of Atlantic Salmon in Nova Scotia and Maine

Salmon spend their early years in fresh water and mature years in the ocean. Atlantic salmon habitat in Nova Scotia rivers has been devastated by increased acidity. A study of 49 rivers that historically supported salmon found populations to be extinct in 14 rivers and severely impacted in 20. Loss of salmon is correlated with increased acidity. Preliminary work suggests that episodic acid deposition has also contributed to the decline of Atlantic salmon in Maine, with the greatest impact occurring in smolts and fry.

Damaged Canadian watersheds, located primarily in southern Ontario and Quebec, have not responded to reductions in sulfate deposition as well or as rapidly as those in less-sensitive regions. At the current sulfur deposition levels, roughly 95,000 Canadian lakes will continue to be damaged by acid deposition.

Acid rain has resulted in large losses of fish and aquatic communities in over 30,000 sensitive lakes in Ontario and Quebec.

In Vermont, 35 lakes have been identified as sensitive and impaired by acidification and declines in base cations in soils.

Forty one percent of lakes in the Adirondack region of New York are either chronically or episodically acidic. The same holds for 15 percent of lakes in New England. Nearly 25 percent of surveyed lakes in the Adirondacks (representative of the range of lakes in the region) do not support any fish, and many others have less aquatic life and species diversity than less acidic lakes. The Catskill Mountains contain many streams with low ANC. Ten years of sampling at four streams indicates a lack of any recovery despite decreases in sulfate deposition and less acid rain.

Reduction in fish diversity in northwest Pennsylvania is linked to aluminum leaching from soils due to acid rain. A comparison of fish data collected in the Allegheny Plateau and Ridge and Valley region 40 years ago to data collected in the mid 1990s found an overall decrease in species diversity, with the most dramatic declines occurring in five species of non-game, acidsensitive fish. Streams experiencing a loss of species had greater increases in acidity and more episodic acidification than streams that either gained or had no change in species.

ACID RAIN GIVES FISH HEART ATTACKS

In fish, the toxic aluminum released from soils due to acid rain disrupts the salt and water balance in the blood. As a result, water moves from blood plasma to the red blood cells, causing the red blood cells to swell and burst. This doubles the thickness of the blood, turning it into the consistency of peanut butter. The fish's heart cannot pump such thick blood, and the fish is deprived of oxygen. If the acid waters don't kill the fish, the toxic aluminum will.

ACID RAIN INJURES FORESTS AND TREES

In places where soils, forests and waters are affected by acidic deposition, the same combination of increased acidity, low alkalinity and more available aluminum can place wildlife at risk both directly and as a result of alterations to food sources and habitats.

Amphibians

Because amphibians require both aquatic and terrestrial environments, they can be particularly susceptible to acid rain induced alterations to habitat and food sources. During the larval stages (just after hatching), aquatic amphibians are most affected by acidic water. Unfortunately, this stage often follows the period of highest water acidity that occurs following snowmelt in the spring. Limited surveys conducted in the eastern United States found lethal levels of pH in waters in 10 to 15 percent of the temporary ponds where sensitive species of amphibians lay their eggs. An additional 10 to 14 percent had pH levels low enough to have other effects such as delayed growth in tadpoles and immune system suppression.

There is also evidence that soil acidification influences the many species of salamander that breed and spend their entire lives in the soil. Not only are eggs affected, but also the distribution and abundance of adults are decreased. A study in New York showed that soil pH influenced the distribution of 11 out of 16 local amphibian species. In the acid-sensitive areas of eastern Canada, more than half of the habitats of 16 of the 17 of the amphibian species there have been affected by acid rain.

Birds

Acidification of the terrestrial and aquatic environments can affect bird populations via a number of pathways, including alterations in diet and food availability, changes to habitat and subsequent impacts on reproduction. For aquatic birds, there is evidence of adverse impacts from acidity to common loons, common merganser, belted kingfisher, osprey, American black duck, common goldeneye, ring-necked duck, eastern kingbird and tree swallow.

Diminished fish stocks from acidic deposition are believed to be playing a role in the impacts occurring to fish-eating birds.

For other terrestrial birds, research undertaken in southern Quebec explored the relationship between changes in forests due to increased acidification and birds. The researchers found that where there was a reduction in canopy cover, the number of canopy birds declined as well.

The Birdhouse Network (TBN), a citizen-science project at Cornell University studying cavity-nesting birds, is undertaking a study to determine whether acid rain is a contributor to the high number of unhatched eggs found in 1999 and in particular to eggs that did not hatch as a result of weak shells.

Earlier research demonstrated a relationship between acid rain, loss of dietary sources of calcium and poor reproduction in birds in the Netherlands. The work suggests that calcium-deficiency, as a result of acidification, may be widespread.

ACID DEPOSITION DAMAGES BUILDINGS

The marble balustrade on the west side of the Capitol building shows damage from acid rain dissolving the mineral calcite. Limestone and marble, the stones used in many buildings and monuments around the world, are especially vulnerable to acid rain. In these stones, strong acids easily dissolve calcium carbonate—the dominant mineral. Many exposed areas on buildings and statues show roughened surfaces, loss of detail in carving and dark streaks. When a sculpture or building is damaged by acid rain, there is no recovery; it is permanently altered.

CLEAN AIR ACT AMENDMENTS OF 1990
WILL NOT SUPPORT RECOVERY

... [T]he Clean Air Act of 1970 and the amendments of 1990 (CAAA) have already been responsible for significant reductions in sulfur dioxide emissions. These reductions have resulted in declines in acidic deposition and some chemical improvements to soils and waters in certain areas of the northeastern US. Surveys show that acidified lakes in the Adirondacks are showing signs of chemical improvement in, what appears to be, response to decreases in acidic inputs. But . . . even with these cuts, emissions remain high when compared to emissions a century ago. Given an accumulation of over 150 years worth of sulfur and nitrogen compounds, it is no surprise that soil build up and subsequent release into waters will continue to damage ecosystems for decades to come.

A number of scientists in North America have grappled with the question of what level of emission reductions will support recovery of ecosystems that have been damaged by acidic deposition. Modeling, combined with long-term monitoring, have yielded the following results:

- The Science Links project of the Hubbard Brook Research Foundation found that an additional 80 percent reduction by 2010 of power sector sulfur dioxide from the level required by the CAAA of 1990 would allow biological recovery to begin mid century in the Northeastern US.

- A modeling analysis conducted for sensitive waterbodies in New York's Adirondacks found that substantial and timely improvements in soil and stream chemistry would occur if acid rain-forming emissions were reduced 80 percent beyond full implementation of CAAA of 1990. Aggressive controls would speed up the chemical recovery in these waters, thus setting the stage for biological recovery.

- Model simulations for Shenandoah National Park project that greater than 70 percent reduction in sulfate deposition (from 1991 levels) would be needed to change stream chemistry such that the number of streams suitable for brook trout viability would increase.

A 70 percent reduction would simply prevent further increase in Virginia stream acidification.

- In the Great Smoky Mountains National Park, application of two separate ecosystem models demonstrated that sulfate reductions of 70 percent are necessary to prevent acidification impacts from increasing at sensitive sites. Deposition reductions above and beyond these amounts are necessary to improve currently degraded aquatic and terrestrial ecosystems.

- At current levels of deposition, analyses at forest sites in the southeastern US suggest that within 80 to 150 years, soil calcium reserves will not be adequate to supply the nutrients needed to support the growth of merchantable timber.

- To reverse and recover from acidic deposition impacts, Canadians in the Acidifying Emissions Task Group have recommended a 75 percent reduction in US power plant sulfur emissions, post CAAA of 1990. Without such a reduction, 76,000 lakes in southeastern Canada will remain damaged.

What Is the Problem With Mercury From Power Plants?

We know that big fish eat little fish. Scientists have shown that when larger fish eat a lot of smaller fish that are contaminated, the contaminants accumulate over time in the larger fish. This is called "bioaccumulation" (*bio* means "life"). So, when humans eat these large fish—for example, tuna—the contaminants in the fish are passed to them.

The following 2003 report from Environmental Defense discusses the issue of mercury in our environment. Mercury is a heavy metal that can spew from power plants and industrial smokestacks. The mercury then settles into oceans, rivers, and lakes. The U.S. Environmental Protection Agency (EPA) recently reported that 8% of all women of childbearing age had enough mercury in their bloodstreams to pose a risk to a developing baby. It is estimated that every year 150 tons of this toxic metal are released into the air. Mercury, like lead, affects the nervous system, and can cause other health problems. In 2004, the Federal Drug Administration (FDA) released an advisory that warned people to avoid eating fish contaminated with mercury. It should be noted that small children or pregnant women are especially susceptible to the dangers of mercury.

In 2003, the EPA began to consider plans to reduce mercury pollution from power plants by 70% over the next 15 years. However, many environmental groups oppose the changes, stating that the plan actually weakens the current rules under the Clean Air Act. In June 2003, 11 states joined together to formally oppose the EPA's proposal that would allow coal-fired power plants to escape the Clean Air Act mandates to reduce mercury emissions. The EPA has decided to review the public comments it received on the plan and to decide whether it will strengthen the rules.

Some states, including Maine and Connecticut, are not waiting for the federal government. They have passed their own regulations against mercury pollution. Although pollution from smokestacks travels across states and countries, Florida has shown that reducing the point sources of pollution within a state can be a very effective way to reduce fish contamination.

The report also notes that the technology to reduce mercury in the environment exists. The time has come, it stresses, to regulate and work with the largest contributors of mercury—the power plants.

—The Editor

Out of Control and Close to Home: Mercury Pollution From Power Plants
by Michael Shore

Mercury pollution from power plants is unregulated and generally uncontrolled. Local mercury emissions in the United States are important contributors to local mercury hot spots, leading to contaminated water, fish that is not healthy for consumption and brain damage in infants.

Mercury is a highly toxic heavy metal that poses a major public health threat. Because mercury can interfere with development, fetuses and children are most at risk. The Centers for Disease Control and Prevention [CDC] estimate that 8% of women of childbearing age in the United States have mercury levels in their blood above what the Environmental Protection Agency (EPA) considers safe. In other words, millions of American women who could be pregnant are exposed to dangerous levels of mercury each year, putting more than 300,000 newborns at risk of brain damage and learning disabilities.

Mercury is released into the air from power plant smoke-stacks and other sources. It can fall to the ground with rain (or without) and enter water bodies in a process known as deposition. People are most often exposed to mercury by eating contaminated fish. The problem of mercury-contaminated fish is widespread, with 43 states issuing advisories to limit consumption of mercury-laden fish. Coal-fired power plants account for about 40% of the mercury emissions in the United States—by far the largest single source. Despite this, no limits exist on mercury pollution from power plants.

FINDINGS

Analysis of emission trends and recent modeling of how mercury is transported and deposited into soil and water leads to three important findings that should influence how policy makers address mercury pollution:

Out of Control
1. Mercury pollution from electric utilities remains completely unregulated.

While other industries have achieved considerable reductions in mercury emissions, mercury pollution from electric utilities is predicted to increase with increased electrical demand. National policies have been successful at reducing mercury emissions from medical waste incinerators and municipal waste incinerators by over 90% since 1990. These sectors provide a model for reductions that could be made in the power plant sector.

Close to Home
2. Mercury pollution within the United States puts fetuses and children at risk.

Since mercury does not break down, it can travel a long way before it is deposited in the environment. However, modeling shows that significant amounts of mercury in waters across the nation come from pollution sources within the United States. Sources in the United States contribute to local mercury "hot spots" and add to global mercury pollution levels, leading to contaminated water, fish that is not healthy for consumption, and brain damage in infants.

3. Local sources can lead to local mercury "hot spots."

Local emissions of mercury are largely responsible for mercury deposition hot spots (locations where mercury deposition is high), providing an excellent opportunity for effective reductions.

Recent modeling suggests that at mercury hot spots pollution sources within the state can account for large portions of the deposition. At hot spots across the United States, local sources often account for 50% to 80% of the mercury deposition. . . . [F]or example, local pollution sources account for over 60% of the deposition in hot spots in Michigan, Maryland, Florida, and Illinois. In another recent analysis in south Florida, dramatic reductions in mercury pollution from local incinerators was accompanied by a lowering of mercury concentrations in large mouth bass by 60–75%, indicating the importance of controlling local sources to reduce local contamination.

RECOMMENDATIONS

Reducing power plant pollution is critical to lowering local mercury deposition and avoiding the dangerous contamination of fish, wildlife and people. EPA is required by the federal Clean Air Act to lower mercury air pollution from power plants. To protect public health and the environment from harmful mercury emissions, state and federal policy makers should take the following steps:

- EPA should issue strong mercury standards for power plants to reduce mercury pollution from 48 tons today to about 5 tons, or a 90% reduction. These reductions are consistent with national standards for other source sectors and achievable through available pollution-control technology.

- States with mercury deposition hot spots should pursue their own mercury pollution standards to protect local water bodies and public health, and all states should press for rigorous national standards.

INTRODUCTION

Mercury is a highly toxic heavy metal that poses a major public health threat. It is released into the air from power plant smokestacks and other sources. It can fall to the ground with rain and enter water bodies in a process known as deposition. Mercury makes more surface waters in the United States unsafe for fishing than any other toxic contaminant, and people are most often exposed to mercury by eating contaminated fish.

The form of mercury found in fish, methylmercury, is a neurotoxin that causes brain and nervous system damage. Even with fish-consumption advisories, exposure to mercury-contaminated fish is high.

Although coal-fired power plants account for about 40% of mercury emissions in the United States—by far the largest source—mercury pollution from this sector remains completely uncontrolled.

The EPA is obligated to propose rules in December 2003, to be finalized one year later, to reduce mercury and other air toxics from power plants. Although other sectors, such as waste incinerators, have already reduced mercury pollution by 90%, power companies have called for much weaker standards. Some power companies argue that mercury pollution is a global problem and national standards would not significantly reduce deposition within the United States. Such an argument is based on averaging deposition across the entire nation, which can be misleading.

Along with reviewing health issues and emission trends, this report examines the available scientific evidence on the local deposition of mercury pollution. Cutting-edge scientific research shows that a significant portion of mercury pollution is deposited locally and regionally, which underscores the importance of strong national mercury standards for power plants.

PUBLIC HEALTH THREATS OF MERCURY

Mercury is one of the most poisonous forms of air pollution.

First emitted into the air as a metal, mercury settles in the beds of rivers, lakes and streams, where bacteria convert it to methylmercury, a highly toxic compound. Methylmercury builds up or bioaccumulates in the bodies of animals, so fish at the top of the aquatic food chain, such as pike, bass, shark and swordfish, may contain mercury concentrations 1 to 10 million times greater than the surrounding water. People are exposed to unsafe levels of methylmercury by eating contaminated fish.

Pregnant women, women of childbearing age, children, subsistence fishers, recreational anglers and Native Americans who consume large amounts of fish are most at risk for health problems caused by mercury exposure. Pregnant and nursing women who eat mercury-contaminated fish place their fetuses at risk for brain damage or other birth defects. The Centers for Disease Control estimate that 8% of women of childbearing age nationally have mercury in their blood streams beyond the levels that the EPA considers safe. Thus, millions of American women of childbearing age are over-exposed to mercury through consumption of contaminated fish, putting over 300,000 newborns at risk of brain damage and learning disabilities each year.

Recognizing the increasing health threats from mercury pollution, the United States Food and Drug Administration [FDA] and 43 states warn against eating several species of fish such as pike, bass, shark, swordfish and mackerel. . . . The geographic extent of areas under mercury advisories increased by almost 138% from 1993 to 2002, with the most dramatic increases having occurred in the last several years.

The increase in fish advisories is not necessarily an indication that the problem of mercury-contaminated fish is worsening, but the increase does reflect that scientists and public health officials have gained an increased understanding of the severity of the mercury-deposition problem. Increased testing of fish for mercury contamination has revealed more species and more water bodies with high mercury concentrations.

COAL-FIRED POWER PLANTS REMAIN
UNREGULATED FOR MERCURY

In 1999, mercury emissions from coal-fired power plants accounted for about 48 tons, or 41%, of new mercury emissions to the atmosphere from the major sources. While the other two largest sources of mercury pollution have declined, mercury pollution from power plants has remained static. In 1990, municipal waste incinerators, medical waste incinerators and power plants were the three largest sources of mercury pollution. Since then, federal regulations have required the clean-up of both medical and municipal waste incinerators, resulting in a 90% reduction in pollution levels.

This leaves power plants as the predominant source of mercury pollution and, of the three largest sources, the only one that is not regulated. The more than 90% reduction from medical and municipal waste incinerators provides a benchmark.

STATES ARE TAKING THE LEAD IN
CONTROLLING MERCURY POLLUTION

In 2003, Connecticut became the first state in the country to regulate mercury emissions from coal-burning power plants. The Connecticut law requires coalfired power plants to achieve either an emissions standard of 0.6 pounds of mercury per trillion Btu [British thermal units], or a 90% efficiency in technology installed to control mercury emissions. According to the company affected by the legislation, PSEG Power, applying the Connecticut standard nationally could cut mercury emissions from power plants up to 86%.

Other states are also considering mercury standards. Massachusetts has proposed a standard to capture 95% of mercury contained in the combusted coal, while Wisconsin's final proposed rule would require an 80% capture efficiency, based on the mercury content of the coal. New Hampshire intends to propose mercury emissions caps on the power sector in 2004. Illinois is evaluating the need for state standards, and North Carolina is reviewing options for reducing mercury pollution.

TIGHT CONTROLS ON POWER PLANT MERCURY POLLUTION ARE NECESSARY AND FEASIBLE

EPA is well aware of the public health threats of mercury pollution. In 1997, the agency presented a comprehensive report to Congress on mercury pollution. In December of 2000, EPA made the determination that it would develop mercury standards for power plants, identifying mercury "as the hazardous air pollutant of greatest concern to public health from the [electric utility] industry." Proposed standards are due December 2003 with final standards required one year later.

Cost-effective technologies exist to reduce mercury emissions by more than 90%, providing EPA the opportunity to develop strong standards.

THE SIGNIFICANCE OF LOCAL MERCURY DEPOSITION

Mercury does not break down, and it can travel long distances before it is deposited. Some power companies use this as an excuse to oppose strong national mercury limits, claiming most mercury pollution comes from outside U.S. borders. However, modeling data show that significant portions of mercury deposited in waters across the nation come from within North America, and often deposition is local.

Atmospheric mercury pollution that has reacted and combined with other pollutants tends to deposit locally or regionally, while unreacted mercury (elemental) tends to enter the global atmospheric pool, enabling it to be deposited virtually anywhere in the world. Even where the global sources are major contributors, it is important to recognize that the large global pool of mercury is not naturally occurring.

The global pool is fed by the emissions that result from the combustion of coal in the United States and around the world. The EPA Mercury Study Report to Congress in 1997 estimates that 66% of all of the mercury deposited in the U.S. comes from national sources, and that 34% comes from sources outside of the U.S. On the other hand, recent modeling supported by the Electric Power Resource Institute (EPRI), the research arm of some of the nation's largest power companies, estimates that on average 70% of mercury deposition comes from global sources. However, the average deposition figure is highly misleading. Averaging modeling results drowns out the high local deposition rates in specific locations across the country. For example, a family eating fish from a water body that is downwind from a nearby power plant might not take any comfort in the fact that average deposition from North American sources may only be 30%. For the family, the local power plant may account for the vast majority of the deposition.

This same EPRI analysis also shows that U.S. sources are responsible for more than 60% of the mercury deposition in the Boston–Washington, DC, corridor, an indication of the importance of local and regional sources. At one of the models' selected receptor areas, Pines Lake, New Jersey, 80% of the deposition comes from sources within the United States, showing that regional deposition can be quite high.

The influence of local emission sources is reinforced by state-of-the-art mercury deposition modeling assessments conducted by EPA. This EPA modeling shows that at mercury hot spots (locations where mercury deposition is highest within a state), local emission sources within a state can be the dominant source of deposition. At hot spots, local sources within a state commonly account for 50% to 80% of the mercury deposition. In-state sources contribute more than 50% of the pollution to sites in the top 8 worst hot spot states. Local deposition hot spots are located across the country, and local deposition estimates would likely be even higher if they accounted for pollution sources in nearby states, not just those in-state.

An ambitious analysis of mercury pollution, deposition and fish contamination in Florida provides on-the-ground evidence that corroborates the importance of local sources. Because of tighter standards on medical and municipal waste incinerators that took effect in mid-1992, South Florida's total estimated local emissions of mercury declined by about 93% between 1991 and 2000. During this same period, mercury deposited via rain and other precipitation declined in South Florida by about 25%. Concentrations of mercury in large mouth bass have also decreased significantly, 60–75% since the early 1990s. These data strongly suggest that reducing local mercury pollution will lower concentrations in local water bodies, and in turn reduce contamination in fish and the risk of human exposure.

RECOMMENDATIONS

Reducing power plant pollution is critical to reducing local mercury deposition and avoiding the dangerous contamination of fish, wildlife and people. EPA is required by the federal Clean Air Act to lower mercury pollution from power plants. To protect public health and the environment from harmful mercury emissions, federal and state policy makers should take the following steps:

- EPA should issue strong mercury standards for power plants that reduce mercury pollution from 48 tons today to about 5 tons, or a 90% reduction. These pollution reductions are consistent with national standards for other source sectors and achievable through available pollution-control technology.

- States with mercury deposition hot spots should pursue their own mercury pollution standards to protect local water bodies and public health, and all states should press for rigorous national standards.

CONCLUSION

Sources in the United States contribute to local mercury "hot spots" and add to global mercury pollution levels, leading to contaminated water bodies, fish that is not healthy for consumption and brain damage and learning disabilities in infants.

International Action

The nations of the world recognize the public health threat posed by mercury pollution. The United Nations Environment Program (UNEP) Governing Council urges all countries to identify populations at risk and reduce human-generated mercury releases, and many nations have initiated measures to reduce mercury pollution.

In North America, the U.S. and Canada Great Lakes Water Quality Agreement calls for the elimination of mercury from the Great Lakes. The New England governors and Eastern Canadian premiers adopted a Mercury Action Plan to reduce mercury pollution in that region. The United States and Canada also joined Europe in signing a 1998 Protocol to the Convention on Long-Range Transboundary Air Pollution to reduce mercury emissions below 1990 levels.

The experience of Florida shows that substantially reducing mercury emissions can dramatically lower mercury contamination in fish and reduce human exposure.

To reduce deposition and environmental contamination, the United States needs to clean up its own sources of mercury pollution. National policies have successfully reduced mercury emissions by 90% in both medical and municipal waste incinerators, and the technology exists for power companies to make similar reductions. Despite being the largest single source, mercury pollution from power plants has never been regulated. It is past time for government to set protective but predictable standards for power plant mercury pollution to protect the nation's children from its damaging effects. Leadership by the United States will not only lower mercury deposition and improve public health within the nation's borders, it will also provide a model to other nations for reducing mercury emissions globally.

How Does Dust Pollute Air and Water Around the Globe?

It is natural for soil and dust to get blown around by the wind. The Hawaiian rain forests have long benefited from the nutrients blown in on dusty soils from Asia. And the Amazon rain forests benefit from wind-deposited dust from far away. A problem occurs when human activity upsets this natural balance.

This has happened in the United States. During the Dust Bowl disaster of the 1930s, dust darkened the skies as topsoil blew from Oklahoma to as far away as Washington, D.C. Poor farming practices were a major cause of the tragic devastation. Today, millions of tons of African top soil are blowing away every year. The remaining soil cannot support vegetation—crops, grasses, or forests. The once-fertile land becomes a desert. This process is called desertification. It is not caused by drought, but by humans overfarming, overgrazing, and cutting down forests. The United Nations (UN) estimates that more than 1 billion people are at risk from desertification.[1] The ruined land leads to famine and poverty. The problem is of such concern that the General Assembly of the United Nations has declared 2006 the International Year of Deserts and Desertification.

But the issues of soil erosion go far beyond the environmental and social consequences of degraded land. The following two articles address what happens to the dust-sized particles after they blow away from the land and travel across continents and oceans. The dust itself causes health problems. Dust particles have been found to have pesticides and viruses attached to them. They also carry the bacteria and fungus that may have caused diseases in coral reefs in Florida and the Caribbean Sea. Although the research is not complete, the signs all point to blowing dust as a global problem. Perhaps by addressing the issues of soil degradation and better land management practices, the problem of toxins and global dust can be addressed as well.

—The Editor

1. United Nations Convention to Combat Desertification. *The Secretary General: Message on the World Day to Combat Desertification.* June 2004. Available online at *http://www.unccd.int/publicinfo/statement/annan2004.php.*

2. United Nations General Assembly. Fifty-Eighth Session. Agenda item 94 (b). December 2003. Available online at *http://www.un.org/special-rep/ohrlls/ohrlls/A-C.2-58-L.60.pdf.*

Dust in the Wind: Fallout From Africa May Be Killing Coral Reefs an Ocean Away
by John C. Ryan

Coughing her way downriver on a slow boat to Timbuktu, Ginger Garrison is a little out of her element. As Bozo tribesmen pull catfish from the Niger River and boatmen pole their dugout canoes through the midday gloom, the strong winter wind known as the harmattan lifts clouds of fine red dust into the air, and into the eyes and lungs of people throughout the dry North African region known as the Sahel.

The only breathing difficulty Garrison, a marine ecologist, usually has to worry about is emptying her scuba tank too fast in the gin-clear, bathtub-warm waters of Virgin Islands National Park in the U.S. Virgin Islands. Garrison, a U.S. Geological Survey (USGS) researcher whose work has focused for nearly 20 years on Caribbean coral reefs, has come here to Mali seeking a source of one of the most widespread ecological collapses ever documented.

An ocean away from the Sahel, coral reef ecosystems around the Caribbean are dying, and scientists are beginning to think that dust from Africa is playing a major role in their collapse. Overfishing, sedimentation, and direct damage from boats and divers, among other threats, have combined with pathogens, climate changes, and hurricanes to severely degrade reefs around the region. Diseases and bleaching have decimated once-dominant species like staghorn and elkhorn corals, longspine sea urchins, and sea fans. Few species or sites have recovered, and carpets of algae—flourishing in the aftermath of overfishing and die-offs of sea urchins and other algae-eaters—now dominate many Caribbean reefs.

Yet researchers remain puzzled by the decline of reefs in apparently pristine stretches of the Caribbean, far from the usual suspects behind coral decline. "We really don't understand why this is happening on a regional level, and it's happening not only in areas where there are a lot of people, it's also happening on remote reefs. Why?" asks Garrison.

Ever since Charles Darwin noted "the falling of impalpably fine dust" while crossing the Atlantic during his famous scientific voyage aboard the *Beagle*, seafarers and researchers have observed African particulates far out to sea. But most studies of atmospheric dust have focused on its potential impacts on the global climate. Only recently have researchers begun exploring the possibility that the hundreds of millions of tons of African topsoil blown by prevailing winds to the Caribbean each year might be having direct, harmful effects on ecosystems and people there.

Dust reaching the opposite shore of the Atlantic is nothing new. Haze from the Sahel occasionally reduces visibility and reddens sunsets from Miami to Caracas, and is the source of up to half the particulates in Miami's summertime air. Pre-Columbian pottery in the Bahamas is made of windborne deposits of African clay; orchids and other epiphytes growing in the rainforest canopy of the Amazon depend on African dust for a large share of their nutrients.

Joseph Prospero of the University of Miami has tracked dust falling on Barbados, at the far eastern edge of the Caribbean, since 1965. He discovered a sharp increase in dustfall around 1970, coinciding with the onset of prolonged drought in North Africa.

The changed African climate, combined with widespread overgrazing of livestock and the spread of destructive, often export-oriented farming practices in the Sahel, were sending vastly greater quantities of exposed soil into the sky. In peak years, winds now drop four times more dust on Barbados than they did before 1970. Satellite photos of the largest dust event ever recorded, in February 2000, show a continuous dust bridge connecting Africa and the Americas.

In the late 1990s, Gene Shinn and other researchers with USGS noted that benchmark events in the prolonged, Caribbean-wide decline of coral reefs—like the arrival of coral black band disease in 1973, mass dieoffs of staghorn and elkhorn corals and sea urchins in 1983, and coral bleaching beginning in 1987—occurred during peak dust years.

Researchers have since found a variety of live bacteria and fungus in dust hitting the Caribbean, defying conventional wisdom among microbiologists that microbes could not survive a five-day trip three miles up in the atmosphere. "Swarms of live locusts made it all the way across alive in 1988 and landed in the Windward Islands," Shinn says. "If one-inch grasshoppers can make it, I imagine almost anything can make it." A 2001 study by USGS researchers found that the number of viable fungus and bacteria in Caribbean air is two to three times higher during dust events than during normal weather conditions.

Although the vast majority of diseases afflicting coral have not been identified (beyond descriptions of the symptoms they cause), scientists have linked dust to at least one specific coral-killing microbe. Garriet Smith and colleagues at the University of South Carolina have identified the pathogen behind the mass die-offs of sea fans, the graceful soft corals of the Caribbean, as *Aspergillus sydowii*—a soil fungus that does not reproduce in salt water. In the very first sample of airborne dust from the Virgin Islands that Ginger Garrison sent to Smith, he found live *Aspergillus sydowii* in its pathogenic form, among many other microorganisms. The fungal disease may also enter the sea in local runoff from deforested areas, but dust studies have established African dust storms as its most plausible source on isolated reefs and near small islands with no forests and little runoff.

In addition to carrying living hitchhikers, clouds of African dust bring intense pulses of nutrients like iron and nitrates that may be stimulating harmful algal blooms and the rapid growth of both coral smothering algae and microbes that cause coral diseases.

Microbiologist Hans Paerl of the University of North Carolina calls the dust—composed of aluminum, silicon, iron, phosphates, nitrates, and sulfates—"Geritol for bugs."

The dust is not so healthy for humans, if only because the fine particles irritate the respiratory tract and can lodge themselves deep in lung tissue. Researchers have barely begun looking into the health effects of overseas African dust but already have some provocative findings. For example, they have found pesticides banned for use in the United States mixed in with dust particles too small for human lungs to expel. "When they have locust plagues in Africa, we get chlordane and DDT that we can't use here anymore, but it comes back to us on the wind," Shinn says.

There may be other unhealthy substances adhering to the particles as well: some studies suggest the dust carries high concentrations of beryllium-7, a radioactive isotope that appears to adhere to dust particles as they travel through the atmosphere. While seeking medical care for her respiratory tract infection in Mali's capital of Bamako, Ginger Garrison asked around and found that lung problems are terribly common in Mali during the dust season. After the seasonal floods of the Niger River recede and its banks dry, mud—mixed with raw sewage, human and animal waste, and miscellaneous garbage left behind—turns to dust. "Microbes, synthetic organics, pharmaceuticals, antibiotics, you name it," Garrison explains. "Then the winds come, and it's a perfect avenue to take those substances aloft, often north toward Europe or west toward the United States." She also observed the ubiquitous garbage burning and wonders what carcinogens, endocrine disrupters, or heavy metals from garbage burning might also find their way into the atmosphere with dust. She hopes to set up a second monitoring station near Bamako to look for heavy metals and synthetic chemicals like DDT, in addition to the station she set up in late 2000 for monitoring microbe levels in dust.

Africa is not the only source of dust that affects faraway places. Nutrients from the deserts of northwestern China sustain Hawaiian rainforests growing on weathered soils. Chinese haze has long afflicted residents of Japan and Korea, where the yellow dust, laden with pollutants picked up from Chinese cities it passes over, is called "the gate-crasher of Spring." South Korean officials suspect that the dust may have been the source of a recent outbreak of foot-and-mouth disease among cattle along Korea's west coast. Last Spring [2001], Korea suffered through 20 days of unhealthy haze from abroad, the longest yellow dust spell there in 40 years. Chinese dust even caused hazy sunsets around the western United States for several days in April 2000. The Chinese, Japanese, and South Korean governments have launched a program to revegetate dust-generating lands in China, and researchers from around the Pacific Rim have begun intensive studies of Chinese dust and its impacts.

To date, the dust blowing from Africa—unlike Chinese dust—has attracted little attention as a public health issue. The desertification (severe degradation of arid and semi-arid lands) that exacerbates dust formation also has serious economic and human consequences close to home: one in six people in Mali have become environmental refugees, forced to leave their land as it turns to dust. Despite the massive amount of land claimed by expanding desertification each year, the phenomenon receives only infrequent attention, perhaps because the effects seldom seem to transcend international borders. These new studies of well-traveled dust may turn that impression on its head.

Given all the locally generated pollution in the Caribbean, it's understandable that African dust is on few people's radar screens. But reversing the decline of the region's once flourishing underwater ecosystems may be impossible without investing more effort in stabilizing the wind-whipped lands of northern Africa.

"It's just another example of how small the Earth is, and how so many things are interconnected: global processes mixed up with how people live their lives," says Garrison. The mounting evidence of damaging fallout thousands of miles from sources of dust may help convince the rest of the world to pay more attention again to the forgotten, dusty corners of planet Earth. "Maybe we're not quite as isolated as we thought from areas with major health problems," says Garrison. "And maybe we should be more concerned about the welfare of people and the land in these far away places."

Dusted
by Hannah Holmes

Half a world away, an ocean apart, the wind is picking up particles of dust. Why is that our problem?

"I walked out on the deck of my boat and smelled Africa," says marine ecologist Virginia Garrison, who was traveling through the Caribbean at the time. "You know how a smell can totally remind you of something? The smell is warm and dry, and it pricks your nose." Garrison is speaking figuratively, but in fact she was smelling Africa—tiny particles of it that had blown off the distant continent about a week earlier during a massive dust storm. Flowing on the wind in a broad, golden river a couple of miles above the Atlantic, the dust had headed west. And now, much diluted, it was settling. Vast clouds of dust regularly blow out of the Sahara in aerial streams thousands of miles long. During winter months, a single storm might eventually sprinkle a quarter of a million tons of dust over the Amazon. In summer, when the dust in northern Africa is boiling up almost constantly, the rivers tend to undulate due west, toward the Caribbean and—a few times each summer—toward Florida. But, while Africa has probably tossed desert dust across the Atlantic for millions of years, today's dust is different.

"People ask, 'If this dust has always come over, what's changed?'" says Garrison, a member of a new U.S. Geological Survey (USGS) team formed to study the hazards of incoming dust. What has changed is northern Africa itself. Parched by drought, heavily farmed by a hungry population, and chemically altered by the use of plastics, pesticides, and other industrial substances, the region is offering different dusts to the wind.

Last year, USGS microbiologist Dale Griffin conducted the first microbial census of African dust trapped over the Caribbean. His results rattled a pillar of microbiology. Traditionally, scientists did not expect many microbes to survive long in the air, where the organisms are assaulted by oxygen and ultraviolet radiation. Only a few microbiologists had speculated that microorganisms could travel long distances. "When I came into this project, I thought if I got three or four organisms I'd be pretty lucky," Griffin says. "Oh, my God, it was surprising!"

No one had predicted the bouquets of pink, yellow, orange, and gray that bloomed in Griffin's petri dishes: "A hundred and twenty species of fungi and bacteria," he says. Griffin thinks the dust may actually protect the microbes from the elements, sheltering them in microscopic crevices like tiny caves. The particles at the top of a dust cloud may also shade those traveling beneath them.

These biological hitchhikers may be the result of a crucial change in the airborne rivers of African dust. Before the 1970s, mineral chips from the Sahara probably dominated the dust that crossed the Atlantic. But in the 1970s and 1980s drought gripped the Sahel, the region south of the Sahara. More than a million people died. Drought and livestock erased vegetation, and Lake Chad shrank, exposing fine sediments. The survivors of the Sahel have had to farm harder, forcing food out of the fragile land and cutting trees to clear farmland and fuel cooking fires, and desertification has taken hold. Dust is one by-product of all this change. Dust from Lake Chad and the Sahel, unrestrained by roots or water, has come billowing west.

Compared with the Sahara, the Sahelian soils are rich in natural fungi, bacteria, and other microorganisms. Griffin believes his 120 species are only the fringes of what's actually there. "If there's one growing on the plate," he explains, "there are probably ninety-nine more that won't grow on a common medium." Of the microbes he did find, one in four is a plant pathogen. A plague of *Aspergillus sydowii* fungus that maimed and killed sea-fan corals throughout Floridian and Caribbean reefs in the 1980s may have flown in on dust from Africa.

Griffin has yet to sieve the dust for viruses—which could outnumber bacteria by a hundred to one. Most viruses are harmless, but a few agricultural bad guys are on his most-wanted list. "We're looking for the really nasty ones," he says. "We'll probably go after foot-and-mouth this summer. And we're looking for citrus canker." This scabbing bacterial disease turned up in Miami in 1995. Hoping to prevent it from reaching central Florida's fruit groves, state crews began culling both commercial and backyard trees within the known blowing distance of infected ones. But with more than $200 million spent and about 2 million trees destroyed, the scourge persists, and Griffin suspects the disease is an import. Citrus canker, he points out, is endemic to northern Africa.

But the sheer volume of dust would cause trouble even if it were pathogen-free. The Sahel catastrophe opened up such vast new sources of dust to the wind that, when the flow peaked in the mid-1980s, the amount trapped by a sampling station on Barbados had quadrupled. Today, dust storms in the Caribbean can still be heavy enough to foul laundry and cars. A dust storm once delivered a swarm of locusts to the region. And now, African dust has been linked to red tide outbreaks.

For decades, scientists have theorized that iron-rich dust might spark plankton blooms. In 1999, Jason Lenes, a graduate student at the University of South Florida, analyzed the fallout of an African dust cloud that settled into the shallow waters off Florida's west coast. Iron in the water shot up 300 percent. An explosion of a plantlike ocean bacterium, *Trichodesmium*, ensued. Then, organic nitrogen, a by-product of the *Trichodesmium* bloom, rose 300 percent. And the nitrogen served as a power lunch for a notorious red alga, *Karenia brevis*.

A single red tide can kill hundreds of manatees and millions of fish and cause respiratory ills in swimmers. This bloom was no exception. "It was a particularly nasty year," Lenes recalls. "There were millions of dollars' worth of damage." He is now trying to determine what percentage of Florida's red tides are fired up by dust. Florida's big *Trichodesmium* blooms almost always follow the arrival of African dust, he says, but how often a red tide follows *Trichodesmium* isn't known.

Lenes hopes his work will help Florida cut its losses. "If you could predict when a red tide is coming," he says, "you could close beaches and fisheries and prevent the waste of harvesting shellfish that are toxic. You could save millions of dollars."

Virginia Garrison recently visited the nation of Mali, which straddles the Sahara and the Sahel, to set up an air-monitoring station. "It's the worst air I've ever breathed, including Jakarta's," she says. "I developed a really bad sinus problem, and it hurt my lungs to breathe." In her travels, she saw impoverished farmers burning garbage in their fields. The burning is a traditional technique for fertilizing poor soils—but today's garbage is increasingly rich in plastics and other pollutants.

Africa is modernizing, and so is its dust. "Now you have pesticide use," Garrison says. "Pharmaceutical use. Plastics. When you burn plastics you get phthalates, which are endocrine disrupters. And all these chemicals may adsorb onto the tiny dust particles." Her dust monitor in Mali, and another in the Virgin Islands, are designed partly to investigate the dust's chemical content.

African dust is already known to carry radioactive beryllium, which forms naturally in the atmosphere and probably builds up as the dust travels. "We couldn't believe how high the beryllium-7 was" in the Virgin Island samples, says USGS geologist Gene Shinn. "One sample was three times the upper limit for the workplace." Then there's radioactive lead, a product of the natural decay of radon in rocks. The dust also ferries toxic mercury in concentrations a thousand times higher than are typical in U.S. soils.

"If dust is contaminated with this stuff, then it's a direct delivery to the lungs," Garrison warns. "And [it]'ll stay right there in your lungs." The average African dust particle is a hundredth of a hair's width in diameter, perfect for lodging deep in human lung tissue. But how much of this dust people are actually breathing in the Caribbean and in Florida—where African dustfalls are much lighter than those in the Caribbean—and how it may affect them aren't known. Some doctors and scientists suspect a correlation between dust and asthma, but the research is just beginning.

Even in its early stages, however, dust research has demonstrated how the most remote environmental tragedy can ripple across a great distance. If scientists continue to link expensive U.S. problems to African dust, then the United States might eventually recognize the benefits of helping to heal the Sahel. "Florida's dust problem is one more proof that Americans can't turn their backs on the rest of the world and hope to live untouched in our castle," says James Gustave Speth, former head of the UN Development Program and an NRDC [Natural Resources Defense Council] trustee. "We have ignored the suffering in the Sahel, and it is to our own peril."

BIBLIOGRAPHY

American Heart Association. *Air Pollution, Heart Disease, and Stroke.* Available online at *http://www.americanheart.org/presenter.jthml?identifier=4419.*

American Lung Association. *State of the Air 2004.* Available online at *http://lungaction.org/reports/.*

American Lung Association and Environmental Defense. *Closing the Diesel Divide.* 2003. Available online at *http://www.environmentaldefense.org/documents/2738_DieselDivide.pdf.*

Clean Air Task Force. *Unfinished Business: Why the Acid Rain Problem Is Not Solved.* 2001. Available online at *http://www.catf.us/publications/reports/Acid_Rain_Report.pdf.*

Eisele, Kimi. "Is Air Pollution Making Us Sick?" *On Earth.* Winter 2003. Available online at *http://www.nrdc.org/onearth/03win/asthma1.asp.*

Gutt, Elissa, et al. *Building on Thirty Years of Clean Air Success.* Environmental Defense Report. 2000. Available online at *http://www.environmentaldefense.org/documents/398_CAAReport.PDF.*

Holmes, Hannah. "Dusted." *On Earth.* Spring 2002. Available online at *http://www.nrdc.org/onearth/02spr/dust.asp.*

Ryan, John C. "Dust in the Wind: Fallout From Africa May Be Killing Coral Reefs an Ocean Away." *World Watch Institute.* January–February 2002. Available online at *http://www.worldwatch.org/pubs/mag/2002/151/.*

Shore, Michael. *Out of Control and Close to Home: Mercury Pollution From Power Plants.* Environmental Defense Report. 2003. Available online at *http://www.environmentaldefense.org/documents/3370_MercuryPowerPlants.pdf.*

Turco, Richard P. *Earth Under Siege: From Air Pollution to Global Change.* New York: Oxford University Press, 2002.

U.S. Environmental Protection Agency. *Pollution Factsheets.* Available online at *http://www.epa.gov.*

FURTHER READING

Dowie, Mark. *Losing Ground: American Environmentalism at the Close of the Twentieth Century.* Cambridge, MA: MIT Press, 1995

Goddish, Thad. *Air Pollution.* Boca Raton: Lewis Publishers, 2004.

Gutt, Elissa, et al. *Building on Thirty Years of Clean Air Success.* Environmental Defense Report. 2000. Available online at *http://www.environmentaldefense.org/documents/398_CAAReport.PDF.*

Turco, Richard P. *Earth Under Siege: From Air Pollution to Global Change.* New York: Oxford University Press, 2002.

WEBSITES

American Heart Association
http://www.americanheart.org/

American Lung Association
http://lungaction.org/

Environmental Defense
http://www.environmentaldefense.org/

Natural Resources Defense Council
http://www.nrdc.org

World Watch Institute
http://www.worldwatch.org/

Acid deposition. *See* acid
rain
Acidification, 90
Acid Neutralizing
Capacity (ANC), 90
Acid neutralizing capacity
of soils, 88–90
Acid rain, xv, 85–95
and aquatic systems,
90–92
forest impact, 92–93
formation, 16, 20, 21, 85
ongoing problems,
86–87, 94–95
pH of, 87–88
regulations, 85, 86
and soil chemistry,
88–90
Aesthetic damage, 19, 22,
93–94
African dust
chemical content, 112,
117–118
in coral reef decline,
109–114
microbial content, 111,
114–116
nutrient content,
111–112, 116–117
Agricultural losses from
ozone, 15, 66–67
AHA (American Heart
Association), 30
*Air and Rain: The
Beginnings of a
Chemical Climatology*
(Smith), 85
Air pollutants
common, 11, 12
indoor, 30, 33
reduction in, 63
toxic, 25–27
Air pollution, ix–x, xvi
in cities, 42–45, 71–73,
75–78, 80–83
exposure levels in
United States, xvi,
3–4, 8

history of, 74–84
in national parks, 63,
64, 69–71
natural sources, 74, 76
Air quality improvement,
steps for, 5–6
Air Quality Index, 9–10,
30, 34–35
Air quality standards,
11, 12, 34–35. *See also
specific pollutants*
ALA (American Lung
Association), 2, 3
Aluminum, 88, 92
American Cancer Society
cohort study, 35
American Heart
Association (AHA), 30
American Lung
Association (ALA), 2,
3
Ammonia, in smog
production, 67
Amphibians, acid rain
and, 92–93
ANC (Acid Neutralizing
Capacity), 90
Aquatic systems. *See also*
Water quality
acid rain and, 90–92
lead contamination,
25
mercury contamination,
100
Aspergillus sydowii, 111,
116
Asthma, 37–46
air pollution and,
40–44
diesel exhaust and,
44–45, 55
immune system in,
39–40, 44–45
incidence, 4
ozone and, 66
smog and, 57
Asthma Awareness
Month, 37

Atlantic salmon, 91–92
Atmospheric deposition,
19
Automobiles, 6, 22–23

Bacteria in global dust,
111, 114–116
Bald eagle, xii–xiii
Bioaccumulation, 96
Biodiversity, viii, xiv
Biodiversity (Wilson),
xxiii
Birdhouse Network, The
(TBN), 92–93
Birds, acid rain and,
92–93
Bronchitis, chronic, 5
Bronx, New York, 43
Brown cloud pollution,
71–73
Browner, Carol, x
*Building on 30 Years of
Clean Air Act Success*
(Gutt), 63
Buildings, damage to.
See Aesthetic damage
Bunker fuel, 59

Canada, 86, 106
Carbon monoxide (CO),
22–24, 33–34, 63
Carcinogens, 25–27,
50–54
Cardiovascular disease,
5, 23, 25, 30–36, 55
Caribbean dust fallout,
109–114
Carson, Rachel, xvii, xxi
Central nervous system
effects
carbon monoxide, 23
lead, 25
mercury, 101
Children
air pollution and, 4
asthma in, 39, 41–46
diesel exhaust and,
44–45, 55

INDEX

lead exposure, 24
mercury exposure, 97
ozone risks to, 65–66
toxins exposure, 27
China, xiii, 113
CITES (Convention on
International Trade in
Endangered Species),
xiii, xxiv
Cities
brown cloud pollution
in, 71–73
health conditions in
children, 42–45
history of pollution in,
76–78, 80–83
pollution control in,
75–76
Clean Air Act, xvi
acid rain, 85, 86,
94–95
mercury contamina-
tion, 96
revised, 11, 86
success of, 63
threats to, 2, 5
Clean Air Nonroad
Diesel Rule, 48
Clean Air Rules (2004),
xvi, 30, 48
Clean Air Task Force,
85
Clean Water Act, xix
Climate change, x–xi,
xvi–xvii, 17
Coal burning
and air pollution,
76–78, 81, 84, 89
environmental damage
from, xv
Coal-fired power plants,
5–6, 8, 20, 102. See also
Power plants
Coffee, shade-grown, xii
Conservation, xii,
xvii–xviii, xx–xxiv
Conservation organiza-
tions, xxii

Convention on
International Trade in
Endangered Species
(CITES), xiii, xxiv
Coral reef decline,
109–114
Crop loss, 15, 66–67

Davies, William Henry,
80
Deaths, excess (defined),
84
Deaths from air pollution,
32, 80–81
Denver, Colorado, 72–73
Deposition
acid. See acid rain
atmospheric, 19
Desertification, 108,
115
des Voeux, Harold
Antoine, 81
Diesel exhaust, 48–62
acute and chronic
effects, 54
asthma and, 44–45, 55
carcinogens in, 50–54
constituents of, 49–50,
51, 57–58
from marine engines,
48–49, 58–62
nonroad use, 6, 48,
57–58
particle pollution in,
54–55, 56–57
regulations, 48–49
Diversity of Life, The
(Wilson), xiv
Dollard, James B., 79
Dubois, Rene, xvii
Dust Bowl, 108
Dust pollution. See
global dust

Earth Under Siege
(Turco), 74
Eastern Acid Rain
Program, 86

Ecosystem damage. See
environmental impact
Elderly, risks to, 4, 55, 65
Eldredge, Niles, ix
Electric utilities. See
power plants
Ellis, Gerry, xiii
Emission control systems,
xi, 23
Emphysema, 5
Endangered Species Act,
xx
Environmental Defense
and Earthjustice, 11,
96
Environmental impact
acid rain, 90–93
carbon monoxide, 23
dust pollution,
109–114
lead, 25
nitrogen oxides, 16–17
ozone, 14–15, 66–67
particulate matter,
18–19, 56–57
sulfur dioxide, 21–22
toxic air pollutants, 26
Environmental move-
ment, xxi–xxii
Environmental policy,
x–xi, xii–xiii, xviii–xix
Environmental Protec-
tion Agency (EPA)
air quality report, 2
air quality standards,
34–35
cardiovascular disease
study, 30
Clean Air Act revision,
11
creation of, xvii, xxii
diesel exhaust standards,
48
on mercury contami-
nation, 96
particle pollution stan-
dards, 68
Eutrophication, 17, 57

Evelyn, John, 77–78, 84
Everglades, viii
Excess deaths (defined), 84
Extinction of species, viii–ix, xiv

Federal Drug Administration (FDA), 96
Fetuses, mercury risk to, 97, 101
Fish
  acid rain effects, 91–92
  lead damage to, 25
  mercury contamination, 96, 98, 101
Florida Everglades, viii
Ford, William, xi
Forests, acid rain and, 92–93
Fossil fuels, xv, 89
Fuel, for ships, 59
Fuller, Buckminster, 80, 84
Fumifugium (Evelyn), 77–78
Fungi in global dust, 111, 114–116

Garrison, Ginger, 109, 111, 114, 117
Gasoline emissions, 6
Gibbs, Lois Marie, xxii
Global dust, 108–118
  chemical content, 112, 117–118
  coral reef decline, 109–114
  microbial content, 111, 114–116
  nutrient content, 111–112, 116–117
Global warming, x–xi, xvi–xvii, 17
Great Lakes Water Quality Agreement, 106

Greenhouse gasses, x–xi, xv, xvi–xvii, 17
Griffin, Dale, 115

Haagen-Smit, Arie, 82–83, 84
Habitat loss, xx
Hazardous air pollutants. See toxic chemicals
Health Effects Institute, 32
Health impact
  carbon monoxide, 23
  diesel exhaust, 44–45, 50–56
  lead, 25
  mercury, 97, 100–101
  nitrates, 68–69
  nitrogen oxides, 16–17, 57
  ozone, 14, 57, 64–66
  particulate matter, 18–19, 54–56, 68–69
  sulfur dioxide, 21
  toxic air pollutants, 26
Heart disease. See cardiovascular disease

Immune system, in asthma, 39–40, 44–45
IMPROVE (Interagency Monitoring of Protected Visual Environments), 68–69
Inconvenience of the Air and Smoke of London Dissipated, The (Evelyn), 77–78
Indoor air pollution, 30, 33
Industrial Revolution, 78–80
Infants, mercury risk to, 97, 101
Inland waterways, 60
Inner-city neighborhoods, and asthma, 42–45

Interagency Monitoring of Protected Visual Environments (IMPROVE), 68–69
International conservation efforts, xxiii–xxiv
International Monetary Fund, xxiv
International Whaling Commission, xxiii

Korea, dust problems, 113
Kyoto Protocol, x, xix

Lakes, acid rain and, 90–92
Lead, 24–25, 63
Lenes, Jason, 116–117
Leopold, Aldo, xxi, xxiv
London smog, 74, 80–81
Los Angeles (photochemical) smog, 74, 82–83

Maimonides, Moses, 77, 83–84
Malakoff, David, x
Marine engines, 48–49, 58–62. See also diesel exhaust
MARPOL, 59
MATES-II (Multiple Air Toxics Exposure Study), 50
Matthiessen, Peter, ix
Mercury contamination, 96–106
  health effects, 97, 100–101
  local deposition of, 103–105
  recommendations, 99, 105–106
  regulations, 96–97, 102–103, 106
  sources, 97–100, 102

INDEX

Methylmercury, 101
Microbial content of global dust, 111, 114–116
Montreal Protocol, xxiii–xxiv
Morris, William, 79
Motor vehicles, 6, 22–23
Muir, John, xiv–xv
Multiple Air Toxics Exposure Study (MATES-II), 50

NAAQS. *See* National Ambient Air Quality Standards
National Air Toxics Assessment, 27
National Ambient Air Quality Standards (NAAQS), 34, 56, 63, 66–67
National Morbidity, Mortality and Air Pollution Study (NMMAPS), 54–55
National parks, xii, 63, 64, 69–71
Natural sources of air pollution, 74, 76
Neutralizing capacity of soils, 88–90
Nicotine, 33–34
Nitrates, 68–73
Nitrogen dioxide (NO_2), 15, 34
Nitrogen oxides (NOx)
  health and environmental impact of, 15–17, 57
  marine engine production of, 57, 60
  in ozone production, 2, 13, 16, 64, 67
NMMAPS (National Morbidity, Mortality and Air Pollution Study), 54–55

Nonroad diesel engines, 6, 48, 57–58
Nutrient content of global dust, 111–112, 116–117

Ozone, ground level
  detector for, 83
  exposure levels in United States, 3, 8
  health and environmental impact of, 14–15, 57, 64–67
  production of, 2, 12–13, 16
  reduction in, 64
  regions with high levels, 64
  standard for, 9–10
  transport of, 14
  trends in, 7–8
Ozone, in stratosphere, 13
Ozonometer, 83

Panda, xiii
Particle pollution. *See also* dust pollution
  deaths from, 32
  health effects, 18–19, 54–56, 68–69
  production of, 17, 67
  reduction in, 63
  regional report, 7
  short-term levels, 3, 9
  standard for, 10, 68
  year-round levels, 3–4, 9
Particulate matter (PM), 2, 17–19. *See also* particle pollution
Passenger pigeon, ix
Pesticides, xi
Phoenix, Arizona, 41–43, 72
Photochemical smog, 74, 82–83

Plants, ozone damage to, 14–15, 66–67
PM. *See* particulate matter
PM2.5, 7, 10
Pope, Carl, xii
Ports, 59–60, 61
Postel, Sandra, xx
Power plants, 5–6, 8, 20, 96–99, 102
Primary standards, 12
Prospero, Joseph, 110

Red tide, 116–117
Residual fuel, 59
Respiratory conditions, 5, 21, 37–46. *See also* asthma
Risk groups for air pollution, 4–5
Roosevelt, Theodore, xvii

Sahel drought, 109, 110, 115–116
Salmon, Atlantic, 91–92
Santiago, Chile, 75–76
Schonbein, Christian, 83
Sculptures, damage to. *See* aesthetic damage
Secondary standards, 12
Shakespeare, William, 78
Shelley, Percy Bysshe, 79, 84
Shinn, Gene, 110
Shipping. *See* marine engines
Silent Spring (Carson), xvii, xxi
Smith, Robert A., 85
Smog
  deaths from, 32, 80–81
  defined, 11
  history of, 76–80
  London, 74, 80–81
  photochemical (Los Angeles), 74, 82–83
  production of, 16, 24, 37, 64

Smoke-generated (London) smog, 74, 80–81

Smoking, 33–34

Soil chemistry, acid rain and, 88–90

Soot, 19

Species extinction, viii–ix, xiv

State of the Air: 2004 (ALA), 3–10

Steam engine, 78

Strategic Ignorance (Pope), xii

Streams, acid rain and, 90–92

Stroke. *See* cardiovascular disease

Sulfur dioxide (SO_2) deaths from, 32
health and environmental impact of, 20–22
in particle pollution, 21, 68
in photochemical smog, 82
reduction in, 63

sources, 19–20, 57–58

Sulfur oxides (SOx), 59, 81

Sunlight, in ozone production, 13, 16

TBN (The Birdhouse Network), 92–93

Tobacco smoke, second-hand, 33

Tokyo, Japan, 44

Toxic chemicals, 17, 25–27, 112, 117–118

Toxics Release Inventory, 27

Trash burning, 6

Trichodesmium bloom, 116–117

Turco, Richard P., 74

United Nations Environment Program, 106

Urban air pollution, 71–73, 75–78, 80–83

Utilities. *See* power plants

van Helmont, Jean Baptista, 39

Vehicle emissions, 6, 22–23

Visibility impairment in national parks, 69–71
nitrogen oxides and, 17
particulate matter and, 19, 56–57
sulfur dioxide and, 21

Volatile organic compounds (VOCs), 2, 13, 64

Water quality, x, xix–xx, 17, 21–22. *See also* aquatic systems

Whaling, xxiii

Wildlife in America (Matthiessen), ix

Wildlife protection, xx

Wildlife refuges, xii

Wilson, E. O., viii, xii, xiv, xviii, xxiii, xxiv

Wood burning, 6, 76

World Wildlife Fund, xxiv

Wright, Anne, 39–40

ABOUT THE CONTRIBUTORS

YAEL CALHOUN is a graduate of Brown University and received her M.A. in Education and her M.S. in Natural Resources Science. Years of work as an environmental planner have provided her with much experience in environmental issues at the local, state, and federal levels. Currently she is writing books, teaching college, and living with her family at the foot of the Rocky Mountains in Utah.

Since 2001, DAVID SEIDEMAN has served as editor-in-chief of *Audubon* magazine, where he has worked as an editor since 1996. He has also covered the environment on staff as a reporter and editor for *Time*, *The New Republic*, and *National Wildlife*. He is the author of a prize-winning book, *Showdown at Opal Creek*, about the spotted owl conflict in the Northwest.